BACK to EDEN

Gift of the

New Canaan Garden Club

New Canaan Library
CONNECT · DISCOVER · GROW

BACK to EDEN
Landscaping with Native Plants

FRANK W. PORTER

ORANGE *frazer* PRESS
Wilmington, Ohio

ISBN 978-1939710000
Copyright©2013 Frank W. Porter and Orange Frazer Press

No part of this publication may be reproduced in any material form (including photocopying or storing in any medium by electronic means and whether or not transiently or incidentally to some other use of this publication) without the written permission of the copyright holder except in accordance with the provisions of the Copyright, Designs and Patents Act 1988.

Published by:
Orange Frazer Press
P.O. Box 214
Wilmington, OH 45177
Telephone 1.800.852.9332 for price and shipping information.
Website: www.orangefrazer.com
www.orangefrazercustombooks.com

Library of Congress Cataloging-in-Publication Data

Porter, Frank W., 1947-
 Back to Eden : landscaping with native plants / Frank Porter.
   p. cm.
 Landscaping with native plants
 Includes bibliographical references and index.
 ISBN 978-1-939710-00-0 (alk. paper)
 1. Native plant gardening--East (U.S.) 2. Native plants for cultivation--East (U.S.) 3. Endemic plants--East (U.S.) 4. Landscape gardening--East (U.S.) I. Title. II. Title: Landscaping with native plants.

 SB439.24.E3P67 2013
 635.90974--dc23
                     2013004810

Book and cover design: Brittany Lament and Orange Frazer Press
Photos courtesy of Tom Barnes on pages 10, 30, 31, 32, 38, 56, 58, 60, 61, 66, 68, 74, 75, 78, 81, 88, 92, 100, 103; photos courtesy of Harold Kneen on pages 16, 44, 46, 50; photo courtesy of Mark Rose on page 44; photo courtesy of Anne Porter on page 108; photo courtesy of the Oregon Historical Society on page 126; photo courtesy of Smith College Archives, Smith College on page 127; photo courtesy of Minnesota Historical Society on page 128; photo courtesy of Julia Margaret Cameron on page 129.

Printed in Canada
First Edition

This book is dedicated to my mother, Juanita.

ACKNOWLEDGMENTS

For the past twenty-five years, I have been growing and studying native plants of the eastern United States. After retiring as a Native American scholar, I returned to southern Ohio and built a small greenhouse. My intention was to start vegetable and flower seedlings for the garden. Little did I know that I had begun a new journey.

I have been fortunate to meet many individuals who graciously shared their knowledge of native plants with me. Peter Heus, proprietor of Enchanter's Garden, challenged me early on to focus exclusively on native plants. We have shared many adventures, including seeing a black panther on Slaty Mountain in West Virginia, as we collected seeds. Deborah Griffith planted the initial seed that ultimately grew to become this book. My good friends Tom Barnes, Harold "Hal" Kneen, and Mark Rose were kind enough to allow me to use some of their photographs. Julie Flannery read and edited the manuscript. Her keen insight vastly improved the text. My wife, Anne, has been more than patient with me. And, finally, I want to thank all of those whose paths crossed mine for their friendship and their help in the writing of this book.

TABLE OF CONTENTS

Foreword by Hope Taft *page xi*
Introduction *page xiii*

The Philosophy

Planting a Sense of Place — 1 — *page three*
Lost Habitats — 2 — *page nine*
Planting Seeds of Hope — 3 — *page twelve*
The Silent Invasion — 4 — *page fifteen*

The Guide

Establishing a Native Plant Garden Using Nursery Grown Plants — 5 — *page twenty-one*
Got Shade? Native Grasses, Sedges, and Rushes for Landscaping — 6 — *page twenty-five*
Gardening Problem Spots — 7 — *page twenty-nine*
Selecting a Groundcover — 8 — *page thirty-six*
Native Trees to Know and Grow — 9 — *page forty-five*
Native Shrubs: Practical Beauty — 10 — *page fifty-one*
Native Vines: A to Z (Almost) — 11 — *page fifty-seven*
Blazing Stars in the Garden — 12 — *page sixty-three*
Gold in the Garden — 13 — *page sixty-nine*
Another Garden Star — 14 — *page seventy-three*
Milkweeds — 15 — *page seventy-seven*
Lobelia: The "l'Obel" Garden Prize — 16 — *page eighty-five*
Rock Gardening with Native Plants — 17 — *page ninety-one*
Hypertufa and Native Plants — 18 — *page ninety-six*
Grass-Like Plants — 19 — *page one hundred one*
Creating a Native Prairie — 20 — *page one hundred five*

The Forgotten Pollinators 21 *page one hundred nine*
Creating a Rain Garden 22 *page one hundred thirteen*

The History

The Shale Barrens of Central Appalachia 23 *page one hundred twenty-one*
Unsung Heroines in the Movement to Preserve Native Plants 24 *page one hundred twenty-five*
E. Lucy Braun and the Relic Prairies of Adams County, Ohio 25 *page one hundred thirty-one*
Stewardship: Who Will Take Responsibility? 26 *page one hundred thirty-six*
Some Thoughts About the Future 27 *page one hundred forty-five*
Bibliographical Essay 28 *page one hundred forty-seven*
About the Author *page one hundred fifty-seven*
Flower Tables *page one hundred fifty-eight*
References *page one hundred sixty*
Index *page one hundred sixty-one*

FOREWORD

Back to Eden: Landscaping with Native Plants is a wonderful personal perspective on why using native plants in your gardening efforts and working for their survival is so important. Although Dr. Porter's window is in southern Ohio and his view is of the Allegheny foothills in Ohio and West Virginia, his advice and wisdom on native plants is universal. He is a knowledgeable plants man.

Dr. Porter has taken up a cause worth digging into because the relationship of plants and their pollinators is vital to our survival.

He infuses just enough scientific fact to give you pause and ammunition to educate others on the importance of a balanced ecology based on native species.

He takes up where wildflower identification books leave off. Read his descriptions of plants and what habitats they like the best and then design your garden spaces to include some of them. This book is full of how-to knowledge learned the hard way.

Learn about the early pioneers in native plant preservation and be inspired to take up the cause.

Frank explores ways to go beyond your own backyard and take the message to others in and out of government. He has done the spadework. Now it is our turn to till the soil, plant the seeds, and grow a greener world that takes less energy, chemicals, water and care once native plants are established. This step-by-step guide is a master plan on what to do.

—Hope Taft, *Former First Lady of Ohio*

Miscanthus sinensis has engulfed the guardrails along a West Virginia highway. A better choice would have been to use *Saccharum alopecuroides*.

INTRODUCTION

We are living in a time of social, economic, and political upheaval. The very fabric of our society seems to be unraveling. In the midst of this turmoil, our precious natural areas and resources—the often forgotten riches that made this country what it is today—are being destroyed by growing numbers of invasive species and pillaged by avaricious politicians and businessmen. Natural areas and the wildlife and plants that occupy them are our heritage. When they are destroyed, they can never be replaced. These threats are not new. Every generation of Americans has witnessed efforts to mine, to graze, to develop, and to exploit our public domain. In 1922, Gene Stratton-Porter cautioned that, "If we do not want our land to dry up and blow away, we must replace at least part of our lost trees. We must save every brook and stream and lake…and those of us who see the vision and most keenly feel the need must furnish the motor power for those less responsible. Work must be done." The time has come once again for all of us to pick up our shovels, rakes, and hoes in order to save our natural heritage for the generations to come.

Getting people to recognize and to use native plants in the landscape has proven to be difficult. There are several factors that help to explain this situation. First and foremost, most of the general public is unaware of the identity and diversity of the native wildflowers and grasses prevalent in their locales. All too often, their firsthand experience with these plants comes from seeing them grow along the roadside of interstate highways and country roads. Some will unwittingly dig these plants from the wild and attempt to grow them in their gardens. When the plants fail to survive the shock of being uprooted from their habitats, the gardener is left with the mistaken impression that native plants are too difficult or finicky to grow. It is so much easier and convenient, they perceive, to purchase non-native, and often invasive, plants from local garden centers.

How can we overcome these mistaken impressions and the unethical means of taking plants from the wild? Educating the public about the benefits of using native species in home gardens or in the cultural landscape is the single most important step to take. And there is a receptive audience. Local garden clubs and Master Gardener programs are just two examples of groups of people who can influence others to learn more about native plants. Garden clubs are always looking for speakers for their meetings, and the Master Gardener program has a well-developed curriculum that fully discusses the use of native plants. As these avid gardeners are made aware of the benefits of using native plants, they in turn are in a position to spread that knowledge.

What can native plant researchers, growers, and enthusiasts tell these folks? One critical piece of information concerns the devastating impact that non-native invasive species is having on our ecosystems. Kudzu (the plant that ate the South and is now devouring the North), Purple Loosestrife, English Ivy, Periwinkle, Russian Olive, Multiflora Rose, Johnson Grass, and Japanese Stilt Grass—to name just a few— are destroying federal and state managed lands, infesting agricultural croplands, and spreading at an exponential rate along our highways and waterways. As these alien species compete with native plant populations, they eventually alter the balance of nature. Wildlife populations dependent on these native plants for food, nesting sites, and cover begin to decline and in many instances to disappear. Another major issue is the dependence of non-native species on irrigation, fertilizer, and pesticides to survive in the garden. This places not only a heavy financial cost on the gardener, but an even greater burden on the environment. Precious water is lost because of excessive irrigation. And the water that is not used by the plants finds its way into rivers and streams, now carrying toxic levels of nitrogen, potassium and phosphate residue from commercial fertilizers, not to mention a myriad of other lethal chemicals. Replacing these dependent non-native species with native plants can minimize the use of water for irrigation and reduce—if not eliminate—dependence on fertilizers, herbicides, and pesticides. Native species have evolved through time to adapt and acclimate to given habitats and ecosystems and grow well without artificial stimulation.

Gardeners and landscapers will quickly realize that native wildflowers and grasses as well as trees, shrubs, and vines are also extremely ornamental, deserving of a place in their gardens. We need to demonstrate that native plants offer a viable alternative to non-native plants in the garden without any loss of beauty in the process. For example, *Buddleia* species are touted as significant butterfly magnets. And so they are. But *Buddleia* is also a plant that has escaped into the wild and now poses a threat to many of our ecosystems. On a recent trip through North Carolina, I saw for myself how this species is spreading uncontrollably into the mountain and piedmont regions. Two excellent native alternatives are species of *Asclepias* and *Liatris*. They, too, are butterfly magnets, but they do not wreak havoc on the environment.

Another example of a persistent invader is *Euonymous fortuneii*, *E. colorata*, and *E. alatus* which are present in yards throughout the eastern United States. They are now also present in fields, on the sides of mountains, and along roadways. Their colorful seeds are spread by birds and other wildlife. Few gardeners are aware that there are three native species of *Euonymous* that offer the same red foliage in the fall and also possess equally distinct seed pods. *Euonymous atropurpureus* var. *atropurpureus*, *E. americanus*, and *E. obovatus* can be used quite effectively in the landscape to introduce native species and eliminate invasive plants.

Native grasses, sedges, and rushes can also be valuable assets to the home gardener. There are a variety of ways these plants can be used. They allow us to explore and better understand the creation of a natural landscape. Although they are perceived to lack the brightly colored flowers of the broad-leaved flora, the subtleties of their form and texture create a lasting beauty that spans the seasons. A natural lawn can pulse with life in a way that heightens the senses and yet also soothes the soul. It also offers an ecologically sound habitat for wildlife that further enriches the gardening experience. Despite these significant attributes, many gardeners have elected to use ornamental grasses from other parts of the world. Why? One answer rests with the development of large-scale nurseries in the United States after World War II. To satisfy the increasing need for plant material to quench the landscaping urges that accompanied the rise of suburbia, they looked for new species from remote corners of the globe. Nurseries turned their attention to those species that exhibited variegated foliage or those that had eye-catching inflorescences, ignoring the possibility that these same plants would or could become invasive.

If you travel Rt. 50 between Parkersburg and Clarksburg, West Virginia you can easily see how invasive these non-native ornamental grasses can become. At first, you notice a single clump of *Miscanthus sinensis*, but before long small colonies appear along the highway. A little further along, there is a two-mile stretch of this grass bordering the guardrails. How did this happen? Look at the

hillside across from this infestation and you will notice the property owner's last name spelled out with clumps of *Miscanthus*. The seeds from these plants have dispersed and now not only crowd the roadside, but have invaded public and state parks in the area. Had the landowner been made aware of the virtues of *Saccharum alopecuroides* (Silver Plumegrass), a beautiful native grass strikingly similar to Miscanthus sinensis, this escape of a noxious grass could have been averted.

The list of native alternatives is almost endless. What is lacking is a comprehensive initiative to get the public, private enterprises, and both state and federal governments to incorporate native plants into the gardens and other landscape projects that are under their authority. Imagine the changes that could take place if government agencies were required to use only plants native to the region? Landscape industries must also be involved in any discussion of using native plants and eradicating invasive species. Many of the invasive species have been introduced through the nursery trade. In fact, some of these species are still offered for sale. The sale of invasive species must be prohibited.

Cardinal flowers growing along Raccoon Creek in Vinton County, Ohio.

PLANTING a Sense of Place

Every plant is native to some place in the world. When humans move it beyond its normal range, it becomes an exotic species. A plant living in its natural range and habitat can grow and reproduce without outside aid. It is in balance and harmony with its surroundings. This plant is neither aggressive nor overpowered. Growing in its own special place, it will always be a thing of great beauty. Enjoying native plants established on your property truly gives one a sense of place.

Native plants exist in intricate communities. This association of plants and animals forms an intricate network of relationships, which is influenced by the soil, hydrology, and climate of the area. Today, there are more than four thousand five hundred species of plants and animals now established in the United States that are of foreign origin.

Although many of these exotic (non-native) species are harmless, approximately fifteen percent are severely altering our environment, causing billions of dollars of damage to agriculture, recreation, forestry, industry, human health, and wildlife habitat. Of the more than three thousand species of plants found in the Ohio Valley, about twenty-five percent are non-native. Nearly sixty of these have invaded our natural areas—woodlands, grasslands, wetlands, and savannas.

As more and more land is cleared for developments, highways, and timbering, we are losing not only our native plants but also the very insects and other wildlife responsible for pollinating specific species of plants. The wanton destruction of plant communities means coincidentally the loss of significant insect life within these communities. Furthermore, the continued and uncontrolled use of pesticides and herbicides kills far too many pollinators. The introduction of new diseases is also taking a toll on pollinators. For example, the population of honeybees has dropped twenty-five percent since 1980. This is of critical importance to us, because pollinators are crucial to one-third of our food supply.

Without specific species of native plants to feed on, their pollinators either die out or move to new habitats. These pollinators have a symbiotic relationship with individual species of plants. We do not fully understand or appreciate this unique relationship between individual species of plants and their pollinators. I have been studying the different species of Asclepias (Milkweeds). The intricacy of how pollination occurs truly amazes me. What saddens me, however, is to see a steady decline in the number of fertile seed pods produced each year.

Planning a landscape for wildlife, whether an entire yard or ten square feet, requires one simply to identify the creatures you want to

Canada lily blooming in my garden seven years after started from seed.

PLANTING A SENSE OF PLACE
Landscaping with Native Plants

attract, to analyze your present landscape, and to design your new one to satisfy the needs of your new tenants. Wildlife needs food, water and shelter. By selecting native plants that will allow for a maximum diversity of flowering and fruiting times, planting them in structural arrangements to provide cover, and creating a reliable water source, you will attract many different animals that might take up residence in your yard. Keep wintering birds in mind as you develop your plan. Learning to identify what wildlife visits your yard can become an enjoyable and educational family endeavor. A good pair of binoculars and a couple of field guides make wonderful holiday presents.

Have you ever asked yourself why you have a manicured lawn and what are the costs to the environment to maintain it? The manicured lawn was a symbol of English aristocracy and wealth. Professional gardeners and groundskeepers kept these lawns neat and tidy. As plants from other parts of the world became available, they soon became prestigious additions to the landscape. After World War II, the rapid development of suburbia in the United States witnessed the appearance of our own version of the manicured lawn and the attendant rise of garden centers to supply non-native grass seeds, exotic flowering plants, and garden tools, machinery and chemicals to maintain the landscaping. Keeping up with the Joneses often meant adhering to neighborhood rules and expectations.

The environmental costs of a manicured lawn are staggering. A grass lawn requires as much as eighteen to twenty gallons of water per square foot. In times of drought, which we have experienced in the Ohio Valley for the last three years, local governments have at times curtailed the use of water on lawns. Even more staggering is that one-seventh of the herbicides and pesticides in the United

States are used to keep these lawns green and healthy. Have you ever stopped to ask yourself why lawn maintenance workers wear protective suits and respiratory masks? How safe are our children who play on these lawns where there are residual chemicals? Then there is the mowing. I am frequently reminded of Tim "the tool man" and his quest for more power. Lawn mowers have become new status symbols—the bigger, the better. In the end, we have become groundskeepers and not gardeners.

The time has come to reduce the size of lawns, lessen our dependence on fossil fuels and machines, minimize noise and air pollution, and create our own sense of place again. Beauty is, after all, in the eye of the beholder. By reducing the amount of lawn, you create space for beneficial gardens that are beneficial to you, your family, and wildlife. How long has it been since you planted a vegetable garden, raised your own herbs, made compost, and truly relaxed outdoors? Maybe it is time to get back to the garden.

Indian Grass prairie along the Scioto River in southern Ohio.

LOST Habitats 2

You are driving down the highway when suddenly you see a glimpse of something bright red growing on the bank beside the road. Too late! By the time you turn your head it is already out of sight. We are such a fast-paced society, always in a hurry to be somewhere. That bright red flower was *Silene virginica*, otherwise known as Fire Pink.

The land we know as Appalachia once possessed rich soils, clear and swift streams, and boundless forests. One early explorer described "timber trees above five foot over, whose trunks are a hundred foot in cleare timber." As other plant enthusiasts ventured into this region, they all "beheld with rapture and astonishment, a world of mountains piled upon mountains." The botanical treasures of this vast Garden of Eden, as it came to be known, attracted the attention of botanists, scholars, and the newly emerging occupation of plant collectors.

Yet the landscape today is only a vestige of what once existed. What did these mountains, valleys, meadows, and streams look like before the "discovery" of the New World? What botanical wonders were found hidden on the mountain slopes and in the deep valleys? Many of the species discovered in what is today West Virginia were endemic to the region. On the slopes of Kate's Mountain in July 1892, John Kunkel Small discovered a new clover, which he later named *Trifolium virginicum* (Kate's Mountain clover). This marvelous plant is now extinct on Kate's Mountain because of development and loss of habitat. Several other new endemic species grew in these amazing shale barrens. These included a new ragwort, *Packera antennariifolia*, and two new species of *Phlox*, *Phlox buckleyi* and *Phlox brittonii*; a large-flowered evening primrose, *Oenothera argillicola*; and a beautiful Leatherflower, *Clematis albicoma*.

Many of these shale barrens are now protected by both state and federal agencies. Other shale barrens are not so fortunate. Without some means of protection, these rare and precious species will meet the same fate as Kate's Mountain Clover.

Silene virginiana

It may come as a surprise, but vast prairies once existed in southern Ohio. Early settlers observed barrens, oak openings, meadows, and prairies where tall grasses dominated the landscape. These grasses were Little Bluestem, Big Bluestem, Indian Grass, Dropseed Grass, and tall, smooth Panic Grass. Some of the early settlers testified that as they rode through these prairies on horseback, they could grasp a handful on each side of the horse and tie them together over their heads. Among the many wildflowers were sunflowers, asters, blazing stars, flowering spurge, purple and yellow coneflowers, Sullivant's Milkweed, Purple Ironweed, and Prairie Dock.

Lynx Prairie

The introduction of the steel plow heralded the destruction of these vast grasslands. The need for farmland, the establishment of settlements, and the construction of roads hastened their disappearance. We will never again be privileged to sit atop a horse and see such a wonder of nature. But we can still experience the thrill of emerging from a deeply wooded area and walking into a remnant prairie.

The Edge of Appalachia Preserve System in Adams County, Ohio is a band of unique nature sanctuaries stretching north from the Ohio River for twelve miles along Ohio Bush Creek. Nearly thirteen thousand acres, which contain the largest list of rare plants and animals in Ohio, are now protected. Other natural areas are also protected in Ohio and West Virginia.

So the next time you are speeding down the highway, slow down and take the opportunity to visit one of these unique preserves and natural areas. It will take you back to an earlier time and place, and allow you to glimpse that former "Garden of Eden."

PLANTING Seeds of Hope 3

American society is once again at a major crossroads. The economy is faltering. Foreign policy has embroiled us in two major wars. Natural resources are declining at an alarming rate. Health care is in desperate straits. And the educational system simply is not working. It takes a catastrophe of one sort or another to awaken us to the perils of facing our way of life. Some hope that the federal government will devise a strategy to remedy the situation. Some choose to play a waiting game, believing that the problems will magically resolve themselves. And others have decided to take matters into their own hands, wanting to return to a simpler time, a more meaningful way of life when family, friends, and church were the foundations to fulfill their dreams.

We have become all too dependent on modern technology. Our every need is satisfied by people, production, and energy beyond our control. As a result, we have become more and more removed from the land that nourishes and sustains us. Today's youth (and many of their parents) do not know how to raise their own food. Many of them have never planted a garden. It is time to return to a greater measure of self-sufficiency, to reinforce family togetherness, to renew our bond with nature, and to strengthen our moral principles.

There is something satisfying about planting a garden, be it a vegetable garden or a flower bed. Gardening puts us in touch with nature. It teaches us to understand and appreciate the fragile ecosystem that supports life on earth. Not so long ago, the North American landscape consisted of vast woodlands, extensive prairies, verdant meadows, grassy savannahs, and wetlands. These natural areas, because of population growth and development, are but vestiges of their original splendor.

Whether you live in the city or the country, your gardening efforts can make a significant difference in helping to restore the natural landscape. Have you ever considered that your garden, no matter the size, could become a wildlife preserve that could help sustain plants and animals that were once common in North America? Have you given any thought to why you chose the plants that are now growing in your garden? In the past, our gardens were created solely for their beauty. They became spaces for us to relax and play. Huge expanses of lawn enveloped the gardens. Our gardens and landscape became a measure of our wealth and social status. Little or no thought was given to the environmental cost of maintaining the way we landscaped our properties.

Natural areas and wildlife preserves became places to visit while on vacation. Automobiles and highways made nature accessible. The loss of habitat and the decline of species nearer to home has been something that we have heard little about. We have it in our power to bring about change by the way we garden. We can once again bring nature close to home.

Using native plants in the landscape is the first critical step in that direction. Native plants have evolved over thousands of years and are adapted to local conditions. Once established, they flourish without the need for fertilizers or pesticides. And only rarely do they require watering. Native plants provide both food and habitat for wildlife, and they contribute to biodiversity. By creating wildflower gardens, we begin to understand what makes our natural areas unique. As wildlife begins to return to our landscape, we also begin to understand how our ecosystem functions. We should be gardening in harmony with nature—not in a battle against nature. We can begin finally to appreciate the beauty and importance of our native plants and wildlife.

Gardening can become a family experience. I recall a spring when my father decided to grow peanuts. My Uncle John happened to be visiting, and he asked my father why he was planting the peanuts with their shells still intact. My father replied: "The shells will quickly rot away and the peanuts will germinate." By mid-summer, not a single peanut had germinated. When queried by my uncle, my father answered: "It must have been a bad batch of peanuts." He never admitted to anyone that the peanuts should have been removed from their shells. I can still see the smile on my uncle's face when he dug a peanut from the ground and found the seed still inside.

How can you begin to create a native plant landscape? You can start by planting a butterfly garden. You can build a rain garden or a small pond that will not only prevent runoff but attract wildlife. You can replace non-native species with native plants in your flowerbeds. You can replace much of your lawn with low-maintenance groundcovers. And you can learn from others and share what you learn. Your native plant landscape will bring years of enjoyment to you and your family. You will have planted seeds of hope for the future.

THE SILENT Invasion 4

A silent invasion is taking place in our precious forests, meadows, and wetlands. Little by little, invasive plants are outcompeting native plants as they vie for nutrients to survive. The list of invasive species is growing at an exponential rate. These invasive plants arrive in cargo containers from abroad either as seeds, roots, or entire plants. They are also brought into this country intentionally by nurseries who sell them to unsuspecting gardeners who are delighted by the flowers and foliage, but are completely unaware of the ecological havoc these plants can cause in our native ecosystems. Attempts to eradicate these unwelcome introductions will always be hampered until the public is made aware of the damage caused by them.

One solution is to begin using native plants as substitutes for these invasive species. Native plants are not only extremely ornamental, they are also well-adapted to the growing conditions in which they will be placed, requiring little or no irrigation, needing no fertilization, and using no insecticides. The use of native plants lessens the destruction of fragile ecosystems that are inundated with chemicals as the result of too much irrigation and use of pesticides and insecticides.

The intentional and accidental introduction of alien species of plants is one of the dire threats to the natural resources of southern Ohio and West Virginia. One need only drive along the West Virginia turnpike and look at the rapid spread of Kudzu vines, Paulownia trees, and the ornamental grass *Miscanthus sinensis* to see how devastating these species can be. Invasive species

Stiltgrass invading a park in Athens, Ohio.

are even more prevalent within the confines of our states forests and parks. *Microstegia vimineum* (Stiltgrass) is literally choking out hundreds of species of native wildflowers and grasses. Stiltgrass is an annual grass that produces thousands of seeds per plant that attach themselves to any object that passes through them. Off road all-terrain vehicles (ATVs) are one of the main culprits. As they ride along trails covered with Stiltgrass, their tires spread the seeds wherever the ATVs venture. All too often, the riders stray off the trails and traverse the sides of mountains or travel along gullies and cuts dissecting the slopes. Within a matter of weeks, there are green strips present where

Ailanthus altissima taking over a woodland area.

the tires have dispersed the seeds. Within three years, Stiltgrass can replace all of the native vegetation on the forest floor.

Tree of Heaven (*Ailanthus altissima*) is not only invasive but also poisonous to humans. There have been instances of individuals who were sawing these trees became seriously ill from the sap in the sawdust. Another pernicious invasive plant that is a public health hazard is *Heracleum manztegazzianum* (Giant Hogweed). Originally from Asia and introduced as an ornamental plant, Giant Hogweed's clear, watery sap has toxins that cause photo dermatitis. Skin contact followed by exposure to sunlight produces painful, burning blisters that can develop into purplish or blackened scars.

Far too many other species have silently invaded our forests, meadows, and waterways. A statewide effort is necessary to prevent and control the continued spread and introduction of these alien species. Efforts are already underway by both federal and state agencies to eradicate specific invasive species. It will prove to be a fruitless effort, however, if these same non-native species continue to grow in adjacent private lands and remain a source of seeds that will ultimately spread back onto public land. The public, as well as federal and state agencies, must be made aware of the ecological catastrophe that is taking place throughout the Ohio Valley because of non-native invasive plants.

Economic development is also taking its toll on native plants. In the past, special interest groups and private companies have ravaged the landscape through widespread timber operations, coal mining, construction of chemical and power plants, highway construction, new homes, and shopping malls. All of these economic ventures have come at the expense of the native flora and fauna. Each year, the number of threatened and endangered species grows at an alarming rate. Sadly, the loss of these plant and animal communities does not often reach the public's attention until long after the destruction has occurred. In fairness, what would the Ohio Valley be without its magnificent mountains, rivers, and streams? The answer will become evident all too soon.

Vinca spreading uncontrollably along a roadside.

What can be done to change how the natural resources of our region are protected and conserved for future generations to enjoy? The answer lies in educating the public and state officials about how diverse and fragile these resources are. Many of our natural resources are finite. When they have been exhausted or exterminated, they cannot be replaced. Other resources may be replenished, but not necessarily in our lifetimes. State officials need to tout these diverse and natural resources as tourist attractions. We have far more to offer than ski resorts and white water rafting. More and more of our natural areas must be set aside and protected. After this is accomplished, there will be opportunities for botanical excursions for tourists as well as locals. The "Rails to Trails" movement, an effort to utilize abandoned railroad beds and make them into trails for hiking or biking, is an excellent example of this process.

Erythronium americanum; Spring ephemerals brighten our woodlands.

ESTABLISHING 5
a Native Plant Garden Using Nursery Grown Plants

Avid gardeners can create a beautiful and fully developed native wildflower garden in one year by installing nursery-grown plants. Several advantages apply to using nursery-grown plants instead of seeds directly sown into the soil. When seeded, wildflowers and grasses typically do not bloom until the third year or even later. These young seedlings must compete with weeds in order to become established. Many plants grown in nurseries, however, will flower the first year due to the fact that they are grown in ideal living conditions and are given time to develop.

The gardener can also place the plants according to a landscape design to create a desired effect. And when plants are used, weeds can be readily distinguished from native plants. It is frequently difficult to differentiate weeds from slow-growing native seedlings.

Native wildflower gardens are a great choice for residential landscapes. By following these simple steps, you can create your own native wildflower garden:

1. The garden area must be completely free of weeds and grasses. The soil should be cultivated to a depth of one foot to break up layers of compacted soil. Do not turn the soil; instead, use a tined fork to create crevices by digging first across the garden and then at four-to six inch intervals. Once you have accomplished this task, add organic matter such as compost, peat moss, well-rotted manure, shredded leaves, and broken down wood chips to a depth of at least four inches. Work this organic material into the top four to six inches of the soil. This will aerate the soil, allow water infiltration, and provide natural nutrients for the plants.

2. Select plants that will flourish in your garden. Use native plants that fit your garden's growing conditions. Do not change the garden to fit the plants. Choose flowers and grasses that match your growing conditions, fulfill your color preferences, extend bloom times, and create texture and depth by employing different heights and types of foliage.

3. Native plants do best when installed in spring or early fall. Early spring flowers often do better when transplanted in the fall. Prepare a map of your garden ahead of time. Space the plants according to height, breadth, bloom time and color. Mark each plant to identify them during the time it takes to establish and to assist with any necessary weeding. Mulch to a depth of four inches with shredded leaves and then cover the leaves with three inches of wood chips. Do not mulch to the stem of the plants; let the mulch gently taper down to the plants. This will reduce disease by allowing air to circulate around the plants. Mulch will reduce weeds, retain soil moisture, maintain uniform soil temperature, and provide nutrients to the plants.

4. Cover native plants with four to six inches of clean mulch. After the plants have gone dormant in late autumn, protect them from soil heaving due to freezing and thawing and winter loss by applying mulch. This is also the time to do a final weeding of the garden and remove all dead foliage from the native plants. In the early spring, remove any excess mulch to encourage new growth. As the new plants emerge, add mulch according to the spring instructions provided above.

Following these simple procedures will ensure a vibrant garden from one year to the next. As you become more familiar with the wide diversity of native plant species, your garden can increase in size accordingly. Winter is the time to read and learn about the native plants of Ohio and West Virginia. As spring approaches, you will be ready to step into the garden and begin to plant.

Elympus virginicus

GOT SHADE? Native Grasses, Sedges, and Rushes for Landscaping

Many home landscapes contain partially, if not predominantly, shaded areas. In the past, gardeners have used Hostas and other non-native plants to fill these areas or they have actually removed existing vegetation to create an environment more conducive to either sun-loving bedding plants or exotic perennials from local garden centers. Native grasses, sedges, and rushes offer an entirely different approach to utilizing these naturally shaded areas. By creating a garden that replicates natural habitats and by selecting varieties of native species that are well adapted to shaded areas, the home gardener can create a landscape that will afford a variety of blooms and foliage that will highlight the garden from early spring through winter. By following some simple procedures, these gardens will require little or no irrigation, fertilizer or pesticides.

A wave of change is flowing through our gardens and landscapes. Garden designers and landscape architectures are increasingly becoming aware of the value and importance of native grasses, sedges, and rushes. One factor that has deterred gardeners and landscapers from using native plants in the past is the mistaken fear that they are invasive and therefore unmanageable. Many of our native grasses, sedges and rushes are truly garden-worthy. It is important to understand that being ornate is not the sole quality of these plants. They have a fundamental beauty and purpose in the landscape.

Native plants offer a variety of ways they can be used in the garden and landscape. Many gardeners hesitate to use native grasses, because they do not understand how to incorporate them into their gardens. Native grasses, sedges, and rushes bring us closer to understanding the diversity of a natural landscape. Grasses are the prevalent species of much of the natural world. They grow in virtually every type of environment, including deserts and Arctic tundra. Native grasses reflect the mood of every landscape. Who among us has not stood in awe as they gazed across a tall grass prairie? Native grasses, sedges, and rushes complement the bright colors of wildflowers, and the subtleties of their different shapes, forms, and textures offer a lasting beauty throughout the year.

Danthonia spicata nestled in the mast of a woodland area.

Native grasses also offer an ease of cultivation. Often they require little more than good soil. There are species that will thrive in full sun, dry shade, moist shade, and wet meadows and bogs. Like any garden plant, they demand only timely weeding and annual grooming. These naturalized lawns require only the simplest means to maintain. In many instances, they will never need mowing. Potential garden spaces on your property should never be forced into submission as manicured lawns.

Elymus species

In the early 1950s, ornamental and native grasses were virtually unused in American gardens. As interest gradually increased in ornamental grasses, plant hunters sought new species from remote corners of the globe. Botanists attempted to raise and introduce new varieties, especially those exhibiting variegated foliage. Although the search worldwide for new and unusual species of ornamental grasses continues to this day, the recognition of the beauty and usefulness of our native grasses, sedges, and rushes in the cultural landscape has lagged woefully behind.

Mounding forms of grasses are of the most importance to us, because they tend to be non-aggressive. Big Bluestem, Little Bluestem, Poverty Grass and Indian Grass transform a lightly shaded area into an inviting sanctuary, especially when intermingled with shade-loving wildflowers.

Sedges are close botanical cousins of the grasses. Sedges comprise approximately one hundred fifteen genera and frequently are found in temperate and arctic regions. Sedges often grow in damp or waterlogged areas, but they also inhabit dry woodlands. Most sedge species are perennial, and many are evergreen, offering a bright contrast to a snow-covered yard.

The Rush family (*Junceae*) is a small one. There are approximately 400 species worldwide. Most rushes grow in damp cool areas. In their flower structure, they are more like lilies than grasses. Species of *Juncus* brighten water features and readily adapt to wet areas where other plants cannot grow.

With the proper choice of species, a gardener can recreate the character of the native sods that existed before the introduction of the modern suburban manicured lawn and utilize shaded areas in new and exciting ways.

GARDENING Problem Spots 7

As spring arrives, you will see many indications that warm weather is approaching. Robins return to mate and search for earthworms. The plants also know that the weather is changing. Trees and shrubs bud out, and some of the perennials show signs of new growth.

This is the time to make preparations for the new growing season. Have you ever had your soil tested to determine its level of acidity or alkalinity? You can purchase a kit at your local garden center or send a soil sample to your extension agent. After you have determined the pH value of your soil, you simply need to pick the best plants for your site.

Ideally, every garden would have the proper soil, sufficient rainfall, and virtually level ground. Unfortunately, most gardens have less than ideal conditions. Regardless of the problem spots in your garden, every gardener can have ample blooms and healthy plants. The secret is to grow wild. Use native plants that are adapted to your area and will grow successfully in the right conditions.

In all likelihood, your garden soil is slightly acidic, having a pH between 6.2 and 6.8. Now is the time to take a walk through your woods or in one of the natural areas in your vicinity and identify acid-loving plants. If you are unfamiliar with how to identify wildflowers, I urge you to obtain a copy of *Newcomb's Wildflower Guide* by Lawrence Newcomb. You will quickly learn to identify wildflowers and become quite familiar with their growing requirements.

Iris virginica

Your next task is to determine any problem spots in your garden or landscape. One of the most frustrating conditions is ground that remains wet throughout the growing season. Rather than going to the expense of having the area drained or filled with topsoil, it makes more sense to select native plants that thrive in these conditions. The choices are extensive. *Iris virginica* (Virginia Iris), *Iris fulva* (Copper Iris), and *Iris versicolor* (Harlequin Blueflag) provide vivid color and will stabilize the soil and will hold back excessive moisture. *Lobelia cardinalis* (Cardinal flower) and *Lobelia siphilitica* (Great Blue Lobelia) offer bright red and blue flowers respectively. *Spiranthes cernus* (Nodding lady's tresses) is a native orchid with white flowers

Iris fulva

Spiranthes cernus

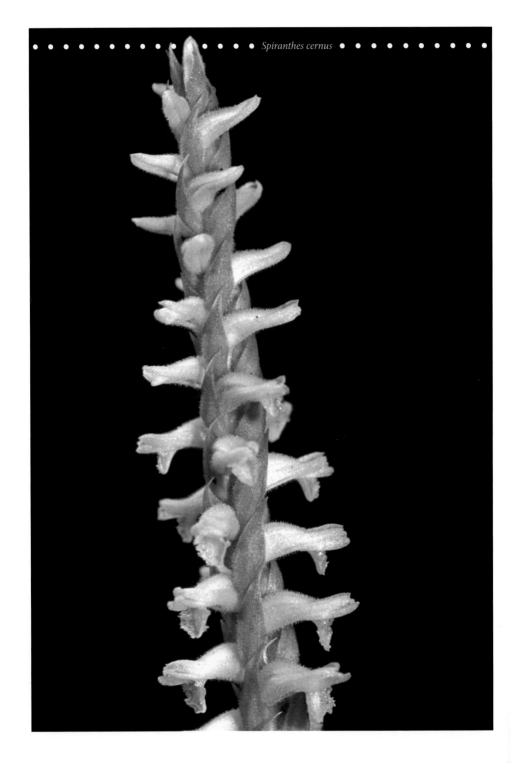

that will colonize a wet area. *Asclepias incarnata* (Swamp Milkweed) enjoys wet feet and will produce pastel pink flowers. If there is sufficient sunlight, *Filipendula rubra* (Queen of the Prairie) will tower over the other plants and produce cotton-candy pink blooms that look like plumy clouds.

At the other end of the spectrum is soil that remains dry. Surprisingly, there are a large number of plants that do quite well without a steady supply of water. For areas that have dry soils and partial shade *Hystrix patula* (Bottlebrush Grass) makes a wonderful backdrop for wildflowers. *Iris cristata* (Dwarf Crested Iris) can serve as a groundcover, especially in and around protruding rocks. *Dodecatheon meadia* (Pride of Ohio) sends up stems whose downward-facing flowers with white petals and gold stamens gives the impression of a meteor shower flaming towards earth. *Silene stellata* (Widowsfrill) reaches above its neighbors and produces deeply fringed white flowers. *Coreopsis verticillata* (Whorled Tickseed) enjoys dry sandy, gravelly or rocky soil and has finely divided leaves with light yellow flowers. *Eurybia macrophyllus* (Large-leaf Aster) and *Symphyotrichum divaricatum* (White Wood Aster) can create beautiful colonies with clusters of large white flowers.

For those of us who live in southern Ohio and West Virginia, hillsides and steep slopes is all too familiar ground and can be difficult places to garden. Spring showers quickly run off, leaving the soil dry and frequently quite thin. Erosion from wind and rain can damage existing plants and even cause mudslides. And yet there are native plants that are equally tough and can meet the challenge of gardening on slopes. Creating a rock garden is one way to handle these growing conditions. Beginning at the bottom of the slope, start to fill in between rocks. *Sedum nevii* (Nevius' Stonecrop), *Sedum*

Iris cristata

glaucophyllum (Cliff Stonecrop), *Hylotelephium telephioides* (Allegheny Stonecrop), and *Sedum ternatum* (Wild Stonecrop) form a mat of low-spreading compact rosettes with flowers that range from white to pink. *Porteranthus stipulata* (American Ipecac) and *Porteranthus trifoliatus* (Bowman's Root) each have bright white flowers with foliage that turns reddish in the fall. Planting these two species together extends the bloom period as the latter flowers two weeks later than the former. *Asclepias quadrifolia* (Four-leaved Milkweed) is the earliest of this species to bloom and has delicate umbels of magenta pink to white flowers. The leaves are in whorls around the stem. To extend the bloom season, plant *Ionactus linariifolius* (Stiff Aster). Its foliage resembles rosemary, and its dark blue flowers add color in late summer and early fall. *Taenidia integerrima* (Yellow Pimpernel) has a delicate appearance with lacy umbels of tiny yellow flowers.

Problem spots do not have to diminish the joy of gardening. Look at them as challenges to enhance the gardening experience. Using native plants is the first step in that direction. Mingle native trees, shrubs, wildflowers, vines and grasses to transform your garden into a small natural area.

Sedum nevii

SELECTING *a Groundcover* 8

One question that I am frequently asked is: "What native plants can I use for a groundcover?" Those unfamiliar with native plants are usually looking for a single species to create a monoculture similar to areas planted with Periwinkle or English Ivy. When using native plants, however, a groundcover should be a diverse mixture of species that not only covers the ground but also creates an attractive and wildlife beneficial area.

The need for a groundcover becomes necessary for areas that are difficult to plant because of the terrain or poor growing conditions. A significant advantage of using native plants for a groundcover is the availability of different species that are already adapted to specific growing conditions. The secret is to select a variety of forbs, grasses, sedges, and even low-growing shrubs to create an ornamental planting.

One way to become familiar with how native plants can be used is to take a hike through diverse areas and note what species are growing in conditions similar to your problem area. Take photographs of the area and then identify as many species as possible. This information can become a model for your groundcover.

The first crucial step in creating a native groundcover is to prepare the soil properly (see Chapter 5). Nothing is more beneficial when amending the soil than applying shredded leaves and compost. In addition, you want to make sure that the newly prepared area does not become a haven for weeds. Because an effective groundcover frequently consists of plants that spread by rhizomes or stolons, you do not want to have unwanted plants, weeds, springing up in the midst of your plants. They not only compete for resources but are extremely difficult to eradicate. An ounce of preparation will alleviate hours of frustration and weeding.

Before you begin planting, decide on the purpose of your groundcover. Are you simply trying to cover up a bare spot and prevent erosion? Are you hoping to encourage wildlife? Are you trying to have flowers throughout the growing season?

Dry and sunny areas can be difficult places to establish a groundcover, but we can take a lesson by studying the plants growing on shale barrens. These shale barrens, which run along some of the slopes of the Appalachian Mountains, present some of the harshest growing conditions plants can encounter. And yet there are several species that thrive in these areas and will do quite well in your landscape. *Cheilanthes lanosa* (Hairy Lipfern), *Cheilanthes eatonii* (Eaton's Lipfern), and *Polypodium virginianum* (Rock Polypody) are small ferns that establish colonies in rocky habitats. When used in your landscape, I would highly recommend placing a few small boulders to enhance the presence of these ferns. The strategic placement of several *Eriogonum allenii* (Shale Barren Buckwheat) will add color, both in foliage and blooms. The dark yellow flowers are emboldened by the oblong, wooly leaves that remain green throughout much of the winter. Another excellent species to add to this mosaic of colors is *Antennaria virginica* (Shale Barren Pussytoes), a low-spreading plant that forms attractive colonies of small rosettes of diminutive silvery leaves topped with white flowers. To add some contrast, include a few clumps of *Allium cernuum* (Nodding Onion). The small, pink flowers will brighten the area. *Hylotelephium telephioides* (Allegheny Stonecrop) can colonize rocky outcrops, and its bright pink flowers in flat-topped clusters will brighten any spot.

Seasonally wet areas can also present some difficult growing conditions. The best approach to creating an effective groundcover is to plant clumps of different species that enjoy having wet feet. *Iris versicolor* (Harlequin Blueflag), *Iris virginica* (Virginia Iris), *Iris lacustris* (Dwarf Lake Iris), and *Iris fulva* (Copper Iris) produce different shades of blue and copper atop strapping green and sword-shaped leaves. *Mertensia virginica* (Virginia Bluebells), with its large ovate leaves and clusters of pale blue flowers that are pink when in bud, appears for a short time in late spring and will colonize open areas along streams. *Equisetum hyemale* (Scouringrush Horsetail)

Mitella diphylla

is an aggressive colonizer and should only be planted in areas that are isolated from your gardens. It does serve a purpose, however, which is to bind the soil in these wet growing conditions. Early settlers also found a use for this plant. Sections of the plant were bound into bundles and then used to scour floors, table tops, and cast iron skillets. *Woodwardia virginica* (Virginia Chainfern) can create a dense and weed-resistant groundcover in moist or saturated soils, especially seeps adjacent to wooded areas.

There are also some excellent choices for areas that are not quite so wet. Bubbling Spring in the Shenandoah National Forest of West Virginia is an intriguing locale. Native Americans once camped beside the spring while on hunting trips. Visitors occasionally find an arrowhead amidst the rocks at the bottom of the spring. *Mitella diphylla* (Twoleaf Miterwort) grows in extensive colonies along the edge of the spring and the banks of the stream that meanders into the woods. When in bloom, there is a profusion of white flowers and on close examination one can see the ring of gold. *Tiarella cordifolia* (Heartleaf Foamflower), with its attractive, feathery spikes of white flowers, makes an excellent companion plant with *Mitella diphylla*. Growing in similar conditions, usually damp and wet woods, is *Phlox divaricata* (Wild Blue Phlox). The color of the blooms is highly variable, ranging from white to pink to blue. In time, it creates a colony with dark green leaves.

Wooded areas with gentle slopes can become a blanket of colors. *Stylophorum diphyllum* (Celandine Poppy) offers bright yellow four-petaled flowers that resemble large buttercups. The silvery foliage brightens these wooded spots, especially as the colony increases in size. *Sedum ternatum* (Woodland Stonecrop), with its dark green leaves and bright white flowers, creates bright patches around the trunks of trees or on mossy logs and rocks. *Packera aurea* (Golden

Ragwort) has violet-shaped leaves that are purple underneath. The yellow daisy-like flowers make it a beautiful early bloomer as it carpets low moist woodland areas.

In areas shaded by deciduous trees, ferns can replace grasses and wildflowers as an effective groundcover. Tall clumping ferns such as *Osmunda cinnamomea* (Cinnamon Fern), *Dryopteris goldiana* (Goldie's Woodfern), and *Osmunda claytoniana* (Interrupted Fern), can be complemented by lower growing and spreading ferns such as *Athyrium filix-femina* (Common Ladyfern), *Dennstaedtia punctilobula* (Eastern Hayscented Fern), and *Thelypteris noveborancensis* (New York Fern). The addition of spring ephemerals can add beauty and color to an already serene setting.

In addition, fern allies can also be used to create effective and ornamental groundcovers. Fern allies have scale-like or awl-shaped evergreen leaves that bear spore cases either in terminal cones or in the leaf axils. *Huperzia lucidula* (Shining Clubmoss) with widely spreading leaves inhabits shady, damp, acidic and humus-rich soil. It is frequently found under dripping sandstone cliffs and along stream bans in wooded areas. *Lycopodium digitatum* (Fan Clubmoss) is stoloniferous, with scale-like leaves. Before the advent of mass produced Christmas greenery, foragers gathered Ground Cedar to satisfy the demand from urban markets for Christmas greenery. This widespread demand reduced the populations of the species in many locales. *Lycopodium obscurum* (Rare Clubmoss) has erect stems about eight inches tall that resemble a miniature tree. It can be found in wet acid soils in rocky wooded areas.

An often overlooked source for a groundcover is native vines. *Aristolochia macrophylla* (Pipevine) and *Aristolochia tomentosa* (Woolly Dutchman's Pipe) are twining vines that normally climb

Cheilanthes lanosa growing in shale barrens near Covington, Virginia.

high into trees. However, they can also clamber over the ground on wooded slopes, effectively creating a dense groundcover. The four-inch long flowers, either yellowish-green or brown-purple, resemble a Dutchman's pipe, giving the plants their names. Both species are also food sources for the Pipevine Butterfly. *Lonicera sempervirens* (Trumpet Honeysuckle), with its bright red and yellow flowers, can also serve as an effective groundcover in a lightly shaded area. *Gelsemium sempervirens* (Evening Trumpetflower) can also be used effectively as a groundcover on a lightly shaded slope. Its profusion of yellow flowers which appear in late spring and again in early fall can enhance an otherwise bland patch of woodlands.

Not to be overlooked are the various species of native grasses, sedges and rushes that can be used effectively to create an attractive groundcover. *Diarrhena americana* (American Beakgrain) is one of the most beautiful and useful plants for a shaded wooded site. Although it is a clumping grass, when planted on one-foot centers it can create a verdant blanket. Its arching stems with green infloresences draw the attention of any passerby. Equally attractive is *Chasmanthium latifolium* (Indian Woodoats). Another clumping form of grass, it too will carpet a wooded area when planted in mass. Its unique seed heads are an excellent companion to Diarrhena americana. Although not quite as striking, *Cinna arundinacea* (Sweet Woodreed) and *Brachyelytrum erectum* (Bearded Shorthusk) form small colonies in a wooded setting. *Bromus kalmii* (Arctic Brome) and *Bromus pubescens* (Hairy Woodland Brome) have distinct infloresences that fit in perfectly with the native grasses. Their arching stems sway gently with the breeze. *Danthonia spicata* (Poverty Oatgrass) can create small colonies around tree trunks in a dry, shaded spot. *Carex pensylvanica* (Pennsylvania Sedge) forms large colonies in wooded areas and can be used effectively to blanket a lightly shaded spot in the landscape. To create a buffer between a wooded and open area, plant *Saccharum alopecuroides* (Silver plumegrass).

Do not hesitate to experiment with different native plants. *Chrysogonum virginianum* (Green and Gold) makes an extremely useful groundcover in either full sun or slightly shaded spots. The common name, Gold Star, certainly suggests the dark yellow flowers that grow above the fuzzy foliage. It spreads quickly to fill in any given area. *Asarum canadense* (Canadian Wildginger), with its heart-shaped leaves and unusual purple-brown flower, certainly draws one's attention in late spring as the wrinkled leaves emerge from the soil. *Hexastylis arifolia var. arifolia* (Littlebrownjug) and *Hexastylis shuttleworthii var. shuttleworthii* (Largeflower Heartleaf) tend to be more evergreen. *Pachysandra procumbens* (Allegheny-spurge) is a far better choice than the non-native species. Although slightly taller, the unique scalloped leaves which are gray-green and the pink to white flowers offer a nice carpet beneath the trees in your yard. Two species that create strawberry-like groundcovers are *Waldsteinia fragarioides* (Appalachian Barren Strawberry) and *Fragaria virginiana* (Virginia Strawberry). The latter has small, edible and sweet berries. And, finally, for those areas that are highly acidic, *Mitchella repens* (Partridgeberry) and *Epigaea repens* (Trailing Arbutus) can be planted beneath evergreens in sandy or rocky woods. The fruit of Partridgeberry is eaten by game birds and some mammals.

Groundcovers should become an integral and complimentary component of your landscape. Forgoing the more traditional monoculture approach for a groundcover, a diverse mixture of native species can not only create year-round enjoyment of foliage and flowers, but also attract beneficial wildlife.

Halesia caroliniana

NATIVE Trees to Know and Grow 9

Native trees should be an integral part of every homeowner's landscape. Sometimes the trees on our property have been adopted from previous owners. Other times, we randomly plant a tree simply because it was on sale at the local garden center and we want to fill an open area. How often do we consider why we need to plant native trees where we live?

There are several very important reasons to incorporate native trees into our home landscapes. Trees are essential to wildlife, because they provide nesting sites for songbirds and offer food and shelter for a wide range of wildlife. Trees can help us to conserve energy. Any landscape without trees can become a "heat island." Without shade, the soil quickly dries out. Imagine the energy and financial savings gained by not having to run air-conditioners constantly during the heat of the summer. Trees also help to clean the air by catching airborne pollutants such as sulfur dioxide, ozone, nitrogen oxide, carbon monoxide, and other particulates. In these uncertain economic times, trees can increase property value in numerous ways. A windbreak can potentially lower heating bills by ten to twenty percent. Some trees can provide edible nuts. Others can be used to create a small orchard. Many flowering species offer berries for wildlife. Trees planted near the street can soften noise pollution. And finally, shade trees planted to the east and west of your home can reduce cooling bills.

Certain species of trees, however, should not be planted in your landscape or should be removed if they are already present. *Ailanthus altissima* (Tree-of-Heaven) came initially to the United States in 1784 as a garden plant. Because it can grow in any habitat except wetlands, Tree-of-Heaven quickly invaded fencerows, roadsides, woodland edges, and our forests. One tree can produce nearly three hundred fifty thousand seeds each year, and seed germination is high. The saplings grow quickly, and the roots give off a toxin that can inhibit the growth of or even kill other plants. Significantly, the sap of Tree-of-Heaven can be dangerous to humans. *Pyrus calleryana* (Bradford or Cleveland Pear) is a very poor selection for the home landscape. Both cultivars are susceptible to damage from heavy ice or high winds. Trees can split down the trunk or large limbs can break off. In addition, the seeds

Cercis canadensis in Kingwood Gardens, Marion County, Ohio.

from these trees escape into other areas and quickly out-compete native flora. *Paulownia tomentosa* (Princess Tree of China) is another fast-growing species and prolific producer of seeds. Anyone who has driven along the West Virginia Turnpike or cruised along the banks of the Ohio River can see how rampantly this tree has spread.

Native trees, especially species that grow in the understory of the forest, make excellent alternatives to these non-native species. *Cercis canadensis* (Eastern Redbud or Judas Tree) possesses a short trunk and rounded crown. In the spring, it is impossible not to notice its leafless twigs covered with purple flowers. These flowers can be fried or eaten as a salad. Its smooth leaves are heart-shaped, and the roots yield a dye. Eastern Redbud prefers moist soil and does well in shaded conditions.

Cornus alternifolia (Alternateleaf Dogwood) is an understory tree with a short trunk and a flat-topped, spreading crown with long horizontal branches. Unlike any other Cornus species, Alternateleaf Dogwood has alternate rather than opposite leaves. The blue-black fruits with red stems provide food for wildlife in the winter. The leaves turn a shade of red and purple in the fall. Although the small white flowers are not as showy as *Cornus florida* (Flowering Dogwood), its dense clusters of flowers, the rich fall color of the leaves and the unusual structure of its branches make Alternateleaf Dogwood an excellent choice for a moist, shaded site. An added bonus is that this species is resistant to anthracnose, a disease that has decimated *Cornus florida* in the wild.

Cornus florida (Flowering Dogwood), with its showy, white flowers, is one of the most recognizable trees in the spring. Unfortunately, anthracnose has devastated this species in the wild. Efforts are underway to develop resistant strains. One solution is to plant trees that have not been exposed to anthracnose. Because it is an airborne disease, do not plant trees taken from the woods or purchase plants from a nursery without assurance they are not diseased. A strong, healthy specimen should give you years of pleasure. Flowering Dogwood enjoys moist soil and some shade. Do not plant one in full sun, because the leaves will be scorched during the summer months.

Cotinus obovatus (American Smoketree) has a short trunk and open crown. The misty flower sprays, resembling puffs of smoke, develop in late spring and remain until fall. Its attractive, brilliant orange autumn foliage makes it an excellent selection for a dry site.

Chionanthus virginicus (White Fringetree) is a small tree with a short trunk and a narrow, oblong crown. Although White Fringetree is one of the last trees to leaf out in the spring, its white flowers in eight inch fleecy clusters are reminiscent of billowy clouds. It does best in rich, moist, and well-drained soil.

The spreading open crown and drooping, bell-shaped flowers of *Halesia carolina* (Carolina Silverbell) make it an excellent choice for a moist, well-drained and shaded spot.

Hamamelis virginiana (American Witchhazel) can reach twenty feet in eight years. Its leaves turn various shades of yellow in autumn. After the leaves drop, the seeds pop out of the fruit pods. Later, the yellow flowers appear, potentially brightening a snow-covered landscape. The bark is used for medicinal purposes. The branches serve as divining rods in attempts to locate underground water. Witchhazel prefers moist, well-drained and shaded areas.

Ptelea trifoliata (Common Hoptree) has a rounded crown. Its bark, leaves, and twigs—when crushed—have a lemon-like fragrance. The flowers have greenish-white petals. The fruit are wafer-like in drooping clusters. At one time, the bitter fruit were used as a substitute for hops. Wafer-ash does best in a moist, well-drained and shaded place.

Euonymous atropurpureus (Eastern Wahoo) inhabits rich woods and stream valleys. In cultivation, it can become a dense, symmetrical, flat-topped tree. The large, five inch leaves turn a subdued red in the fall. The tiny purple flowers, mostly hidden by the leaves, are quite attractive. The pink popcorn fruit capsules, however, that open in the fall to reveal the bright red berries justify its use in the home landscape.

Staphylea trifolia (American Bladdernut) possesses striped bark, compound opposite leaves, and inflated, papery seed capsules. It grows in moist woods and along stream banks. American Bladdernut's stone-like seeds rattle in their pods when ripe.

Unlike the towering oaks and maples found in many yards, the native trees described in this chapter reach between fifteen and twenty-five feet in height. Their foliage, flowers, and fruit can be easily seen and enjoyed. They provide shade for small areas, offer different types of foliage, and produce a wide array of fruit. As an added bonus, native spring flowers can be planted beneath them.

Bradford Pear saplings growing along a highway in North Carolina from seeds dispersed from a nearby nursery.

Purple berries of *Callicarpa americana*.

NATIVE 10
Shrubs: Practical Beauty

Some of the most beautiful native plants in your garden can be shrubs. Whether a shrub is planted as a single specimen or used in mass to create an accent, its different shape, foliage, and color will always create a distinct silhouette in the landscape. Because of the smaller size of shrubs, you will be able to enjoy, up close, their fragrance and blossoms. Many shrubs attract butterflies and birds to your yard; others serve practical purposes such as a hedge or informal border, providing privacy by screening areas from view or creating effective windbreaks and noise buffers.

Native shrubs can meet the growing conditions of every situation in your landscape. The secret is to select the proper species for the right place. One of the more challenging conditions in any landscape is a wet area. Some gardeners attempt to install permanent drains.

ornamental. Black Chokeberry also has white flowers, but blooms later in the spring. The fruit are at first purple, but change to black. Both species provide food for wildlife.

Callicarpa americana (American Beautyberry) makes its contribution to the landscape in late summer and fall. Silvery magenta berries cover the shrub. The flowers are bluish to lavender-pink and are funnel-shaped, clustered in the leaf axils. It is an excellent addition to the butterfly garden. *Calycanthus floridus* (Eastern Sweetshrub) offers an intoxicating fragrance from its maroon (and on occasion greenish-yellow) flowers, which bloom in summer. In earlier times, housewives planted this shrub near the kitchen window to enjoy its fragrant strawberry aroma.

For areas that tend to be dry and receive several hours of sunlight, *Hypericum prolificum* (Shrubby St. John's Wort) and *Hypericum kalmianum* (Kalm's St. John's Wort) are perfect for difficult areas. Shrubby St. John's Wort is a four to six foot mounded shrub with golden yellow flowers appearing in late summer. Kalm's St. John's Wort produces larger and more abundant golden yellow flowers. It also is a mounding plant, but somewhat shorter in stature. *Ceanothus americanus* (New Jersey Tea), which was used as a substitute for English tea during the American Revolution, has creamy white flowers that appear in summer when few other plants are in bloom. It tolerates a wide range of growing conditions so long as the soil is well-drained.

Lindera benzoin (Northern Spicebush) is a shrub for all seasons. In early spring it produces pastel yellow flowers before the leaves appear, reminding one of Forsythia in bloom. During the summer, the bright green leaves droop and exude a fragrance reminiscent of Allspice. In the fall, the shrub is covered with bright red, oval-shaped fruit that are also highly fragrant when crushed.

As is so often the case, native shrubs are under used in the landscape. Homeowners all too often select non-native species, many of which have become highly invasive and obnoxious intruders in our natural areas. Never plant non-native species of *Euonymus* in the landscape. Birds spread the seeds into wooded areas where they germinate and quickly multiply. During a recent trip to the mountains of North Carolina, I noticed entire slopes covered with *Euonymus alatus* (Burning Bush). Few gardeners realize that we have three native species of *Euonymus—atropurpureus, americanus, and obovatus—* that brighten the fall days with similar red foliage. Other non-natives that should be removed from the landscape are *Berberis thunbergii* (Japanese Barberry), *Elaeagnus umbellata* (Autumn Olive), *Elaeagnus angustifolia* (Russian Olive), *Ligustrum ssp.* (Common, Japanese, and Chinese Privet), *Spiraea japonica* (Japanese spiraea), and *Lonicera morrowii, L. tartarica,* and *L. maackii* (Bush Honeysuckles).

Flower of *Aristolochia macrophylla* resembling a Dutchman's pipe.

NATIVE Vines: A to Z (Almost) — 11

Using native vines can be both an exciting and challenging gardening experience. The number of species to choose from is surprisingly large, from the diminutive Allegheny Vine (*Adlumia fungosa*) to the sprawling Woolly Dutchman's Pipe (*Aristolochia tomentosa*). The secret is selecting the right vine for the proper place and understanding completely the growing habits of each species.

Native vines offer three types of growing habits: rambling, twining, and sprawling. In making your selection of species, be sure that your vine does not strangle the tree or shrub that it depends on for support and does not grow to such a length that it becomes unmanageable. Place plants in this category in a natural setting and allow them to grow unattended. Following is a selection of native vines suitable for the garden or landscape.

Adlumia fungosa

Adlumia fungosa (Allegeny Vine) produces white or pinkish flowers that droop in loose clusters from the axils. The leaf stalks twine gently around other plants. It is found on moist ledges and wooded slopes in the mountains. Allegheny Vine is a biennial that will seed itself. Grow it near a shrub so that it can support itself.

Aristolochia macrophylla (Pipevine) is a twining vine that can reach more than thirty feet in length. The dark green leaves are heart-shaped and up to ten inches long. Its yellow-green flowers are reminiscent of a curved pipe. This vine is well adapted to very shaded areas. At one time Pipevine was used to create a living curtain to screen porches from the sun. Today, use Pipevine as a screen on an old wooden fence. *Aristolochia tomentosa* (Woolly Dutchman's Pipe) has a woolly pubescence that coats the new growth. This species is more heat tolerant and requires less shade.

Bignonia capreolata (Crossvine) is a self-clinging vine that can reach more than fifty feet in length and offers striking, trumpet-shaped blooms that are usually reddish brown with a yellow or orange interior. Crossvine supports itself with small, disk-like pads that adhere to any porous surface. Crossvine prefers moist soil and sun. Use it on fences or naturalized in trees.

Campsis radicans (Trumpet Creeper) should not be used in the garden. It clings by numerous roots along the stem and is very difficult to control because of suckering. In a naturalized area, however, its three-inch long, orange trumpet flowers are a welcome sight in moist woods with light shade to sun.

Celastrus scandens (American Bittersweet) is a twining vine that reaches more than fifty feet in length. American Bittersweet will strangle and climb over anything it touches. The red-orange and yellow berries, which require a male and female plant, are its most attractive feature. Use American Bittersweet on a steep bank or in other difficult areas.

Centrosema virginianum (Spurred Butterfly Pea) and *Clitoria mariana* (Atlantic Pigeonwings) are climbing or trailing vines that rarely reach more than five feet in length. Both grow in dry, open woods or barrens. The lavender flowers resemble butterflies in flight and bloom for an extended period. They are a welcome addition to any garden trellis.

Clematis viorna (Vasevine) is a climbing perennial that reaches twenty feet. Although lacking petals, the sepals form a leathery red calyx with a white lip. Found in wet woods, it is at home on a trellis or trailing along a fence. *Clematis virginiana* (Devil's Darning Needles), with showy white flowers, climbs on trees and fences at the edge of woods and fields. This vine is too aggressive for the garden and should be grown in a naturalized area.

Centrosema virginiana

Gelsemium sempervirens (Evening Trumpetflower) possesses daffodil yellow trumpets of flowers that bloom early in spring and again in late summer. Trellises, arbors, and fences in dappled shade offer excellent support for this fragrant vine that can reach twenty feet in length.

Lonicera sempervirens (Trumpet Honeysuckle), with its orange, red and yellow flowers, blooms from late spring to early winter. It is an excellent source of nectar for hummingbirds. Grow it on a trellis or fence in sun or light shade.

Clitoria mariana

Matelea carolinensis (Maroon Carolina Milkvine) has dark maroon flowers and heart-shaped leaves. Closely related but growing further north is *Matelea obliqua*. It also has similar dark maroon flowers and heart-shaped leaves, but the flowers extend out from the stem on a pedicle. These twining vines grow in rich woods and thickets. They will fit nicely on a trellis or fence.

Menispermum canadense (Common Moonseed) is a slender, twining vine that reaches twenty feet in length. It has

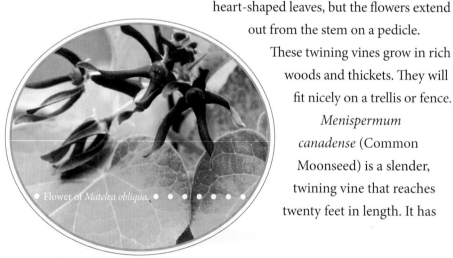
Flower of *Matelea obliqua*.

Flower of *Clematis viorna*.

dark green leaves with greenish-yellow to white flowers that open in early summer. The bluish-black drupes contain moon-shaped seeds.

Passionflowers, *Passiflora incarnata* (Purple Passionflower) and *Passiflora lutea* (Yellow Passionflower), create a tropical garden environment. Both species are climbing or trailing. They have stolons that can easily become invasive. Plant them in a wooded area and enjoy their marvelous flowers.

Strophostyles umbellata (Pink Fuzzybean) is a ten-foot trailing or climbing vine that is frequently overlooked in dry open woods. Its bright pink flowers standout when grown on a trellis or fence.

Wisteria frutescens (American Wisteria) produces blue-purple flowers with a yellow spot in late spring to early summer. American Wisteria, unlike its Asian cousin, is a well-behaved vine that will trail along a split rail fence and produce blooms for nearly a month.

Do not hesitate to incorporate native vines into your garden design. By making the right choices, you can enhance the range of colors and the height and breadth of your garden.

Matelea obliqua beginning to twine onto shrubs at Lynx Prairie in Adams County, Ohio.

BLAZING Stars in the Garden 12

Species of Liatris, commonly known as Blazing Star or Gayfeathers, offer some of the most striking blooms during the hot and dry months of summer. All of the *Liatris* species are perennials that grow from a thickened rootstock. Their leaves are alternate and entire. *Liatris* flowers are unusual because they bloom from the top of the inflorescence down instead of from the bottom up. The showy inflorescence exhibits intense shades from a light purple to a rose-purple to white.

In the wild, each species of *Liatris* inhabits a different ecosystem. Some thrive in moist meadows, while others prefer dry prairies, rocky outcrops, sand dunes, and shale barrens. The one thing they all have in common is the need for at least one-half day of sun.

Native Americans discovered various medicinal uses of *Liatris*. The Cherokee, for example, devised a decoction of the root to relieve intestinal gas, increase urination, and ease backache. Folk medicine uses included treatment of sore throats, kidney stones, and gonorrhea. Another unusual use of *Liatris* among Native Americans involved chewing the corm and then blowing it up the nostrils of horses to keep them from getting out of breath.

Nothing is more suited for later summer color in a dry location in your garden than *Liatris*. All of the *Liatris* species are easy to grow. The difficulty arises from which species to choose for your garden. In the southeastern United States, there are an estimated twenty-two species. We can divide these species into two groups. One group develops an inflorescence that is wand-like in appearance. These inflorescences can range from six inches to as much as two feet in length. The other group has inflorescences on which the individual florets are borne on short petioles that appear as buttons or carnation-like flowers.

Liatris spicata (Dense Blazing Star) is the most common species grown by gardeners. Unlike other species of *Liatris*, it prefers moist, open meadows. *Liatris pycnostachya* (Prairie or Cattail Blazing Star) is the tallest of the species, reaching nearly six feet in height. Its florescence can be nearly two feet in length. *Liatris microcephala* (Smallhead Blazing Star) is the smallest of the species, seldom growing more than eighteen inches in height. However, its inflorescence appears like a purple

Liatris pycnostachya rising above the grasses in Lynx Prairie in Adams County, Ohio.

Flowers of *Liatris ligulistylis*.

fountain emerging from the soil. It is perfectly suited for a rock garden or as a border along a path. *Liatris squarrosa* (Scaly Blazing Star) has deep green, shiny, leathery foliage with unique large, well-stalked, button flowers that appear in late summer. *Liatris aspera* (Tall Blazing Star) possesses long flower spikes with each producing a one-inch lavender-tasseled flower. *Liatris cylindracea* (Ontario Blazing Star) may have few individual flowers (some even solitary) but nevertheless has numerous florets.

Flower of *Liatris cylindracea*.

The bracts are tightly appressed in a cylinder. It is an excellent candidate for rocky, open woods, bluffs, and prairies. And it prefers alkaline soil. *Liatris ligulistylis* (Rocky Mountain Blazing Star) produces crimson-red buds that open to reveal brilliant purple-pink flowers. It is unsurpassed in attracting Monarch butterflies.

Some *Liatris* are rare and distinct from the previously named species. *Liatris hellerii* (Heller's Blazing Star) is a federally endangered species that only grows in a very specific habitat. This species was probably never common due to its restricted and isolated habitat requirements. Because there are so few sites left, Heller's Blazing Star has become vulnerable to seemingly minor threats such as trampling by hikers, climbers, and sightseers. Heller's Blazing Star produces one or more flowering stalks, which rise above a rosette of narrow basal leaves and culminate in a stalk of lavender flowers. *Liatris mucronata*

(Cusp Blazing Star) has long, needle-like leaves that resemble a bottlebrush and has attractive purple flowers. *Liatris scariosa var. nieuwlandii* (Nieuwland's Blazing Star) is a very rare species found growing on the southern shores of the Great Lakes. With pink flowers and wide leaves, it makes a striking impression in the garden. *Liatris odoratissima* (Deer's Tongue or Vanilla Plant) does not possess a corm but rather develops fibrous roots. It flowers in September and October. The leaves, when dry, have a pleasant vanilla odor.

With so many species of *Liatris* from which to choose, gardeners can create a panorama of color, height, and foliage to brighten their summer and fall gardens. One way to maximize the effect of *Liatris* is to create a tiered garden design in which the taller species are placed in the back. The middle area would contain the species reaching between two and three feet, and the front would be a border of the more diminutive species, such as *Liatris microcephala*. As an added bonus, this small garden would become a haven for butterflies.

Flowers of *Solidago caesia*.

GOLD in the Garden
13

The abundant spring rain brings forth a rainbow of colors in our native plant gardens, especially for those who prepared and planted their beds the previous year. Prairie and meadow flowers are in bud and will begin blooming by early July. But spring is also the time to be thinking about your fall garden. What can you plant to highlight your fall garden? The answer is a few species of *Solidago* (Goldenrods).

Goldenrods (*Solidago* species) are, without a doubt, gold in the garden. Unfortunately, Goldenrods have been given a bad reputation by nurseries and gardeners. Because Goldenrods begin blooming in late summer, they often signal the beginning of hay fever season. Goldenrods *do not* cause hay fever. Their pollen is too heavy to become airborne. Another problem frequently associated with Goldenrods is that they are aggressive spreaders and have no place in the garden.

This description is only partly correct. Some species of Goldenrod are stoloniferous, having horizontal runners that root at each node. These species are best used to naturalize an area of your landscape. Other species, however, are not quite so aggressive and add a special beauty and architecture to the garden.

All Goldenrods are showy perennials with many small flower heads borne in clusters. With only one exception, the ray and disk flowers are yellow. Individual species of Goldenrods can be distinguished by the types of their inflorescence. One group has their flower heads located in small, short-stalked clusters in the axils of the upper leaves. Another group possesses heads that are crowded closely together to form a narrow wand-like terminal inflorescence. The next group offers long panicles that are somewhat recurved and appear as if the flower heads are only on the upper side. And the final group has flat-topped inflorescences. The following species can be used effectively in your native plant garden.

Solidago caesia (Blue-stemmed or Wreath Goldenrod) can be found arching out from a wooded bank with its bluish stems and yellow flowers growing in the leaf axils. *Solidago flexicaulis* (Zigzag Goldenrod) has a zigzag stem and broad oval-toothed leaves. Its flowers are on short stems extending from the leaf axils.

Solidago odora (Anisescented Goldenrod) is a fragrant species whose leaves smell of anise when crushed. Blue Mountain Tea is made from the leaves. The arching branches of its panicles are densely flowered. It prefers dry, sandy soil in full sun to partial shade. *Solidago ulmifolia* (Elmleaf Goldenrod) is similar, but the inflorescence has long spreading branches.

Oligoneuron rigidum var. rigidum (Stiff Goldenrod) has unusually large flower heads and a stout stem covered with fine hairs. It prefers full sun and is often found on prairies or openings in wooded areas.

Solidago rugosa known as "Fireworks," is a compact and clump-forming plant with radiating inflorescence that truly resembles fireworks. I have used it effectively both to create a small border or as an accent plant.

Other species of Goldenrods can overwhelm a garden. An excellent example is *Solidago canadensis* (Canada Goldenrod), which frequently overtakes open fields and gives the mistaken impression that all *Solidago* species are too aggressive for the garden. They are better suited to naturalize an area of your landscape or to help remedy a problem spot.

Solidago speciosa (Showy Goldenrod) lives up to its name. Large pyramidal clusters of yellow flowers sit atop reddish stems with smooth leaves. Wet areas are always difficult places to establish plants, but *Solidago patula* (Roundleaf Goldenrod), *Solidago uliginosa* (Bog Goldenrod), and *Oligoneuron ohioense* (Ohio Goldenrod) thrive in swamps and bogs. They quickly spread to help control runoff water and prevent erosion, while offering beautiful yellow flowers.

For those of you with dry banks or barren spots in full sun, *Solidago bicolor* (White Goldenrod) and *Euthamia gramnifolia* (Grass-leaved Goldenrod) will help to colonize the area and present striking and unusual flowers. Silver-rod is the only species of *Solidago* with cream-colored or white florets.

Euthamia gramnifolia growing in a disturbed area in Zaleski State Forest in Ohio.

Grass-leaved Goldenrod lives up to its name, possessing narrow, lance-shaped leaves and topped with flat clusters of flowers.

The last species I want to discuss has a fairly wide geographic distribution, and it is an excellent candidate for a rock garden or stone terrace. *Solidago rugosa var. aspera* (Cliff Goldenrod) grows on the face of cliffs or rock outcrops and has arching stems with yellow flowers. In the garden it resembles a dwarf Forsythia.

The fall season does not have to be limited to the changing colors of the trees. By combining species of *Solidago* with asters, you can have a wide range of colors in bloom in the fall.

ANOTHER Garden Star 14

The blossoms of native asters brighten roadsides, meadows and prairies in the autumn. Although the flowers of many asters are white, several others range from blue to purple and some exhibit beautiful pink petals. In addition to their ornamental qualities, asters are an important food source for butterflies and are essential to the survival of over wintering colonies of honeybees and other social insects. The seeds that are produced after flowering become food for migrating and resident songbirds.

The genus name comes from the Greek *aster*, meaning star. And stars in the garden they are! While some gardeners consider asters a weedy plant, they simply do not understand their growing requirements and have planted them in the wrong location.

Asters should be an integral part of every native plant garden. Regardless of your growing conditions, there is a species of aster that will fit perfectly into your design. As always, it is a matter of selecting the right plant for the right spot. One important point to remember is that some asters are stoloniferous, which means they produce underground runners and can quickly colonize an area. Other asters do not spread and can be planted comfortably among other species.

For those who have wooded areas or shade gardens, there is a broad selection of asters from which to choose. *Symphyotrichum cordifolia* (Common Blue Wood or Heart-leaved Aster) produces blue-violet to rose-colored flowers and will form an attractive patch in a wooded area. *Eurybia divaricata* (White Wood Aster) offers pure white flowers and will also colonize areas beneath the canopy. *Eurybia schreberi* (Schreber's Aster) also has white flowers, but the leaves are larger. *Eurybia macrophylla* (Bigleaf Aster) has violet or lavender flowers with harsh and thick basal leaves. It is an excellent substitute for Hostas. Each of these asters should be used to naturalize an area and are not good candidates for a cultivated garden.

If you have a moist and shaded spot, *Symphyotrichum prenanthoides* (Crooked-Stem Aster) has blue-violet flowers and leaves that clasp the stem. The stem zigzags and has several flowering

• • Flowers of *Symphyotrichum cordifolia*. • • •

branches. *Symphyotrichum puniceum* (Purple-stemmed Aster) also has blue-violet flowers with clasping leaves and a purplish stem. *Eurybia radula* (Rough-leaved Aster) produces violet flowers with lanced-shape leaves that are coarsely toothed. Because of habitat destruction, this species is becoming endangered.

Flowers of *Symphyotrichum oblongifolium*.

Symphyotrichum novae-angliae (New England Aster) can reach six feet in height. It has violet-purple flowers but on occasion produces bright pink flowers. It can become rather aggressive, so use caution where you plant it. *Doellingeria umbellate* (Parasol Whitetop) has white flowers that grow in a flat cluster.

Several species of asters thrive in dry, wooded areas. *Oclemena acuminata* (Whorled Wood Aster), whose upper leaves are larger than the lower ones, has white or purple-tinged flowers. *Symphyotrichum undulatum* (Waxyleaf Aster) has leaves that clasp the stem and are wavy-margined. The flowers are blue-violet. *Symphyotrichum lowrieanum* (Lowrie's Blue Wood Aster) is unusual in that its leaves are greasy to touch. *Symphyotrichum sagittifolius* (Arrow-leaved Aster) possesses either blue, pink or white flowers and leaves that are arrow-shaped at the base. It is similar to Heart-leaved Aster except that it has winged petioles.

Some aster species thrive in dry, sunny locations. *Symphyotrichum ericoides* (Heath Aster) has white flowers and narrow leaves with rough edges. It will spread slowly, so give it some room. *Symphyotrichum dumosum* (Rice Button Aster) has pale lavender or sometimes white flowers borne on numerous flowering branches. *Ionactis linariifolius* (Stiff Aster) resembles a Rosemary plant. It has bright blue flowers with a yellow disk and appears almost like a small woody shrub. Stiff Aster would do wonderfully in a rock garden. *Symphyotrichum patens* (Late Purple Aster) enjoys rugged conditions. The flowers are numerous and surround a yellow disc, which brighten a fall afternoon. *Symphyotrichum oblongifolium* (Shale or Aromatic Aster) is among the last wildflowers to bloom before winter. Its violet-purple flowers form bushy mounds. *Symphyotrichum laeve* (Smooth Aster) has blue-violet flowers and clasping smooth leaves. *Symphyotrichum lateriflorum* (Calico Aster) has numerous diminutive white flowers usually with a purple disk.

For those of you with brave hearts, you might want to consider planting an entire garden area with the native asters of Ohio and West Virginia. When they begin to bloom in the fall, you would truly have a garden of many colors. The wildlife that would come to visit your garden would be amazing. And unlike fall mums, which frequently do not survive our harsh winters, your asters will return year after year.

MILKWEEDS 15

One of the most breathtaking summer-blooming wildflowers is Butterfly Weed (*Asclepias tuberosa*). Its umbels of orange, yellow, and occasionally red blossoms attract not only hundreds of butterflies but many other insect pollinators. The common name, Milkweed, in all probability stems from the fact that when cut or bruised all parts of the plant exude a white and sticky sap. This milky sap contains poisonous glycosides similar to those found in Foxglove plants. The genus, *Asclepias*, comes from the name of the Greek god of medicine, Aesculapius. Native Americans and early colonists used several *Asclepias* species for medicinal purposes, including removal of warts.

Monarch butterfly caterpillars ingest this milky sap, but they are not harmed by it. Instead, their bodies become bitter tasting and even poisonous to predators. Even after their transformation into beautiful butterflies, this protection continues.

Butterfly Weed is only one of more than twenty species of *Asclepias* growing east of the Mississippi River. The leaves of *Asclepias* species are usually opposite, although sometimes they are whorled and on rare occasions alternate. The leaves vary in shape from thick ovals to very thin needles. The stems frequently are simple and solitary. It is the variations in their flower structure that help to distinguish these species. Every small blossom has a downward-pointing corolla, which consists of five petals and five hidden sepals. The upward-pointing part is a five-lobed crown of hoods with horns around a central column of stamens and pistils. They are connected together by a pedestal. Several blossoms are grouped together to form an umbel.

These showy flowers produce nectar that is attractive to bees and butterflies. The unusual structure of the flowers, however, allows pollination in two complex ways. In early summer, buds begin to develop. Pollen sacs develop on the stamens. Visiting insects must accidentally slide one of their legs through a slit and into the interior of the flower. These pollen sacs snag on the insect's leg and are pulled from the stamens. The insect must successfully remove its leg and the attached pollen sac from the slit. If the insect is not successful, the leg may be left behind or the insect may die,

Flowers of *Asclepias quadrifolia*.

permanently stuck to the flower. I have found the legs of insects on several flowers. If successful, the insect must reach another flower and have the leg containing the pollen sac slide through the slit of the second flower. The pollen must be inserted into slits behind the crown perfectly, or the pollen grains germinate the wrong way and are wasted. Once all of these conditions are satisfied, the Milkweed flower has successfully been pollinated. This is one reason why so few pods develop on most plants.

Milkweed, in addition to being a significant source of nectar for numerous butterfly species, is the host plant for Monarch butterflies. Without milkweed, Monarchs would be unable to survive, and we would no longer have their bright orange and black wings to grace our gardens.

Milkweed has also been useful to humans. The roots of *Asclepias tuberosa* (Butterfly Milkweed) were once used to make a tea or tincture for inflammations of the lung (pleurisy)—hence the common name Pleurisy Root. A poultice made from the roots helped to heal bruises, swellings, and rheumatism. *Asclepias incarnata* (Swamp Milkweed) was also used to make a root tea as a laxative. Early settlers used it to treat asthma, syphilis, worms, and rheumatism and as a heart tonic. *Asclepias quadrifolia* (Fourleaf Milkweed) was used by the Cherokee to make a root tea for kidney stones. The leaves were also rubbed on warts to remove them. *Asclepias syriaca* (Common Milkweed) had similar uses. The silky seed tassels served as stuffing for pillows and feather beds. During World War I it was used to stuff life jackets of sailors and pilots. A word of caution, however, is necessary. All parts of milkweeds are potentially toxic, and the silky seeds are extremely flammable.

The habitats of milkweeds range from sand hills to swamps and from full sun to shade. The most widespread milkweed is *Asclepias tuberosa*. It grows along roadsides and meadows throughout the summer. Butterfly Weed is unlike other species of *Asclepias* because it does not have a milky sap and the leaves are alternate. *Asclepias syriaca* (Common Milkweed) did not originate in the Middle East as its name suggests. Instead, Linnaeus mistakenly placed it in a pile of plants from that region and gave it the species name of *syriaca*. It attracts hordes of pollinators, and it is the primary food source of Monarch butterfly larvae. It is a tall, downy plant with broad oval leaves that bear slightly drooping umbels of pinkish-purple blossoms. It should not be grown in the garden because it has underground stolons that will quickly spread. This characteristic, however, makes it an excellent candidate for restoration of natural areas. *Asclepias speciosa* (Showy Milkweed) is similar to Common Milkweed, but the flowers have larger and more tapering upright petals and an abundance of wooly hairs.

Some species of *Asclepias* prefer marshes, swamps, stream banks, or moist meadows. *Asclepias exaltata* (Poke Milkweed) inhabits moist woods and wooded edges. The white or pinkish flowers are arranged in spreading and frequently drooping clusters. *Asclepias incarnata* (Swamp Milkweed) grows in marshes and shores of wetlands. Clusters of pink to rose-purple (and occasionally white) flowers will brighten any moist

A colony of *Asclepias tuberosa* growing with *Spiranthes cernus* in the Trella Romine Prairie in Marion County, Ohio.

Flowers of *Asclepias verticillata*.

garden area. *Asclepias lanceolata* (Fewflower Milkweed) has a red corolla and orange crown and has narrow lanceolate leaves. It favors savannahs, swamps, and marshes. *Asclepias longifolia* (Longleaf Milkweed) possesses a greenish white corolla with rose tips. Its narrow linear leaves grow in irregular pairs. It is a denizen of savannahs, low pinelands, bogs, and swamps. *Asclepias rubra* (Red Milkweed) has a dull red-lavender corolla and opposite ovate-lanceolate leaves. It grows in bogs, swamps, and wet meadows. *Asclepias perennis* (Aquatic Milkweed) has a white corolla and elliptic-lanceolate leaves with long pedicels and petioles. It inhabits swamp forests.

Other species of *Asclepias* grow in open woods. *Asclepias variegata* (Redwing Milkweed) haunts the shade of woodlands. Its umbels of white flowers have a red-purple ring around the pedestal. These bright white flowers are like beacons in the woods. *Asclepias amplexicaulis* (Clasping Milkweed) has a rosy-pink crown and yellow-green corolla. Its large oblong leaves possess wavy margins and are grown on reddish stems. *Asclepias quadrifolia* (Fourleaf Milkweed) has a greenish white to pink corolla. The flowers grow on long pedicels, and the four leaves are in whorls around the stem. *Asclepias verticillata* (Whorled Milkweed) has small flowers with a greenish white corolla. The narrow and threadlike leaves are in whorls around the stem. *Asclepias purpurascens* (Purple Milkweed) has a deep rose corolla and opposite ovate to elliptic leaves. It is becoming increasingly rare due to loss of habitat.

Several species of *Asclepias* are quite rare and in dire need of conservation and protection. *Asclepias cinerea* (Carolina Milkweed) is a small plant with a lavender corolla and narrow linear leaves. It grows in pine/palmetto flat woods and savannahs. *Asclepias connivens* (Largeflower Milkweed) is quite rare and can be found in wet savannahs. *Asclepias obovata* (Pineland Milkweed) has a yellow-green corolla with opposite elliptic to ovate leaves. It prefers sandy pinelands. *Asclepias pedicellata* (Savannah Milkweed) has a greenish cream corolla and narrow linear leaves. It inhabits savannahs along the coastal plain. *Asclepias tomentosa* (Tuba Milkweed) has a yellow-green corolla with an orange tint and tuba-like hoods. The leaves are oblong. It grows in sand hills and pine woods. The flowers of some milkweeds are greenish. *Asclepias viridiflora* (Green Comet Milkweed) has a pale green corolla and oblong leaves with a wavy margin. Although not abundant, it can be found in fields and along roadsides. *Asclepias viridis* (Green Antelopehorns) has a yellow-green corolla with reddish to pink-purple hoods and oval pointed leaves. It prefers fields and pine woods.

No matter what part of the eastern United States you live in, there are species of *Asclepias* that should be an integral part of your garden or landscape. Milkweeds not only provide a wide array of flowers but also serve a functional role as host plants for certain species of butterflies and other insects. And we must begin to take notice of the rare and increasingly endangered species of Milkweeds and initiate efforts to protect and conserve them so that future generations will have the opportunity to enjoy these beautiful wildflowers.

Flowers of Asclepias purpurascens.

Flowers of *Lobelia siphilitica*.

LOBELIA: The "l'Obel" Garden Prize 16

Matthias de l'Obel, born in Lille, France in 1583 and who died in 1616, moved to England and became physician and botanist to King James I. The plant genus *Lobelia* and the botanical family *Lobeliaceae* are named in his honor. L'Obel never beheld the plants that bear his name, but I am sure that he would appreciate and acknowledge their beauty.

Cardinal Flower (*Lobelia cardinalis*) and Great Blue Lobelia (*Lobelia siphilitica*) are two of our most beautiful wildflowers. Both species inhabit wet areas, especially along stream banks. What many gardeners do not know is that there are several other species of *Lobelia* that can be used successfully in their native plant landscape design. Most of these *Lobelia* species grow in meadows, ditches, swamps, and occasionally on rocky hillsides. They offer a range of colors from the brightest red to the purest white.

Lobelia species are easily identified by their bilaterally symmetric flowers which possess a two-lipped corolla. The upper lip has two lobes, while the lower lip has three spreading lobes. The five stamens are united within a tube. The pollen is at the bottom of the tube. When pollinators (frequently butterflies) visit these *Lobelia* species, the pollen drops onto their backs.

Like so many of our native plants, Native Americans and early settlers found medicinal uses for the *Lobelia* species. A word of caution: *all Lobelia* plants are considered to be toxic. Never use them as a home remedy. *Lobelia inflata* (Indian Tobacco) is a strong emetic, expectorant, and sedative. Native Americans smoked its leaves for asthma, bronchitis, and sore throats. It was also used to induce vomiting, thus its nickname "pukeweed." The roots of *Lobelia siphilitica* were used by Native Americans to make a tea for syphilis. Tea made from the leaves was used for colds, fevers, worms, croup, and nosebleeds. Tea made from *Lobelia spicata* (Pale-spiked Lobelia) was used by Native Americans as an emetic. And a wash made from the stalks was used to treat "bad blood" and neck and jaw sores. *Lobelia cardinalis* had similar properties, but was considered much weaker and therefore was not used widely.

Lobelia cardinalis and *Lobelia siphilitica* are the two most frequently grown flowers in this genus. The common name, Cardinal Flower, refers to the bright red robes worn by Roman Catholic cardinals. It grows throughout the eastern United States and is usually found in wet soil, stream banks, and roadside ditches. The first time you come upon this plant in a half-shaded spot along a stream you will never forget the fiery redness that is accentuated by the shadows. I have grown this plant successfully in good garden loam in full sun, but it never attains the vivid red flowers it bears when in its natural habitat. It is an erect and typically unbranched perennial that normally reaches four feet, making it one of the largest *Lobelia* species and the only one with red flowers. The leaves are alternate, lanceolate (about six inches) with toothed margins. Ruby-throated hummingbirds usually pollinate the bright red flowers, which grow in a spectacular raceme. On rare occasions, you might find a pink or white form. Long-tongued butterflies, such as the Spicebush Swallowtail, Eastern Tiger Swallowtail, Pipevine Swallowtail, Cloudless Sulphur, and Dogface butterflies will also visit the flowers.

Flowers of *Lobelia cardinalis*.

Lobelia siphilitica (Great Blue Lobelia) grows in swamps and along stream banks throughout the eastern United States, west to Wyoming and south to Texas and Georgia. The species name, *siphilitica*, comes from Latin "of syphilis" and refers to the use of the root to treat syphilis. It was an ineffective treatment. Bees pollinate this species. Great Blue Lobelia is a robust and erect perennial that can reach four feet but usually grows two feet in height. The flowers are a bright blue with a touch of white. Although this plant prefers moist soil, I have also grown it in good garden soil in full sun.

Lobelia puberula (Downy Lobelia) possesses a fine hairy stem, with toothed leaves and a one-sided raceme of purplish blue flowers. Unlike *Lobelia siphilitica*, it lacks the inflated tube and striping. Downy

Lobelia enjoys open woods and drier soils, usually growing along the wood's edge in dappled shade. Although it grows erect, it has a tendency to flop over and rest on neighboring plants. Nevertheless, it is a welcome addition to the wildflower garden.

Lobelia spicata (Palespike Lobelia) is found from Alberta and Quebec, south through most of the eastern and mid-western United States and west to eastern Texas. It is an erect and unbranched perennial that reaches two feet. Approximately twenty bluish white flowers are scattered along a nearly leafless stem. A rosette of club-shaped leaves grows at the base of the stem. The fruits are tiny, round capsules filled with chestnut brown seeds. *Lobelia spicata* grows in moist or low prairies, but also inhabits meadows, glades, barrens, and thickets. Like other *Lobelias*, it contains poisonous substances formerly used in medicinal remedies and as an agent to discourage the use of tobacco.

Lobelia inflata (Indian Tobacco) is an annual and can become quite weedy. It is by far the most common of the species and is frequently found in disturbed areas. Indian Tobacco is easily identified because of the inflated seed pods that develop along its stem. The flowers are typically light blue to white. And the stem is long, hairy, and branched. Because of the toxicity of all parts of this plant and its tendency to become weedy, it should not be planted in the garden.

The remaining species

Flowers of *Lobelia spicata*.

of *Lobelia* are extremely rare and endangered or are endemic to certain parts of the eastern United States. These species need to be protected and enjoyed during visits to natural areas. *Lobelia kalmii* (Ontario Lobelia) grows in damp soil usually in calcareous regions in meadows, bogs, and ditches. The flowers are white to blue with a conspicuous white eye-spot. *Lobelia boykinii* (Boykin's Lobelia) is semi-aquatic and grows in swamps and cypress ponds from the coastal plain of Delaware to Florida. The lower portion of the plant is seasonally submerged in water. The flowers are blue to white, and it is the only Lobelia species that possesses rhizomes. Boykin's Lobelia is endangered throughout its entire range. *Lobelia dortmanna* (Dortmann's Cardinal Flower) is strictly an aquatic plant found along the margins of ponds and lakes. The flowers are light blue to white and are widely spaced on long flower stems. The upper stem is leafless, while the lower leaves are under water forming a dense basal rosette. The plant is usually eighteen inches in height. *Lobelia nuttallii* (Nuttall's Lobelia) grows in various moist soils from pine savannahs and pocosin to low woods and prefers sandy soil. The small flowers are blue with a white center and two small green spots at the base of the lower lip. The plant ranges from eight to thirty inches in height. *Lobelia elongata* (Longleaf Lobelia) has a long and wispy stem. The flowers are a deep blue and are confined to one side of the stem. The lower leaves are grass-like with sharply toothed outer margins. It prefers wet habitats such as swamps and wet meadows. A similar species, *Lobelia glandulosa* (Glade Lobelia), is fairly common in the coastal plain and eastern Piedmont. It is distinguished from Longleaf Lobelia by having long hairs on the inside of the lower lip of the corolla. *Lobelia appendiculata* (Pale Lobelia) has a slender, erect stem reaching about two feet. The flowers are pale blue to white. It is widely distributed in pinelands and prairies in the Gulf Coast states. *Lobelia appendiculata var. Gattingeri* (Gattinger's Lobelia), named in honor of Augustin Gattinger, a pioneer botanist from Tennessee, is a diminutive variety reaching only twelve inches. Gattinger's Lobelia is an endemic found only in the cedar glades of the Central Basin of Tennessee, although there is a report of one population in Kentucky.

Folks living in or visiting Florida might have an opportunity to see three rare species of Lobelia. *Lobelia floridana* (Florida Lobelia) is native to Florida's wet flatwoods, cypress pond margins, and bogs, especially in the central and western panhandle. Flower colors range from blue to pink to white. *Lobelia homophylla* (Pineland Lobelia) inhabits pinelands, fencerows, and roadsides, while *Lobelia feayana* (Bay Lobelia) grows mostly in pinelands.

Lobelias should be incorporated into any native wildflower garden or landscape. They provide a vivid range of colors while in bloom, and they attract numerous pollinators to gather their nectar. We also need to take immediate steps to protect all of the species facing loss of habitat, especially those growing in and dependent upon wetland conditions.

ROCK 17
Gardening with Native Plants

Rock gardening is a highly specialized way of gardening with plants normally adapted to high altitude growing conditions. It was at one time the preserve of wealthy landowners, especially in England, who not only had the means to create rock gardens, but also had enough influence to gain access to the rare and difficult-to-grow species to plant in them. Plant collectors searched far and wide to satisfy the demand for these plants. It was not long before gardeners in the United States joined in this rock gardening mania. Specialized nurseries came into existence, offering rare and unusual species from every continent.

The creation of the North American Rock Garden Society in 1934 made it possible for more gardeners to enjoy this unique hobby. In recent years rock gardening has once again grown in popularity.

Pink flowers of *Hylotelephium telephioides*.

While enthusiasts continue to search the globe for plants to add to their collections, many aficionados are surprised to learn that there are beautiful native species growing in the Ohio Valley perfectly suited to use in rock gardens.

Constructing a rock garden does not necessarily have to be an expensive or labor-intensive endeavor. One sure way to ensure success is to start small and learn through trial and error. There are six important steps to follow: plan carefully; prepare the site; be prepared to move the stones; set the stones in position; make up and use the planting mixture; and select the plants and proper mulch.

Another way to enjoy rock gardening is to use troughs, more commonly known as hypertufa containers. Hypertufa containers are planters for alpine species and diminutive plants. Hypertufa is a mixture of Portland cement with peat moss and perlite. It is much lighter than concrete. Growing plants in troughs allows one to cater to the individual needs of different species and to grow many more plants in a restricted space. It also brings you into close proximity with your plants. Imagine, also, the ease of weeding, watering and controlling pests.

What to grow? There are a surprisingly large number of native plants from our region that are suitable for use in rock gardens or hypertufa containers. Three species of Sedums are essential to any rock garden. *Sedum ternatum* (Woodland Stonecrop) is a low-sprawling perennial that grows on rocks, logs, or even bare soil and prefers dappled shade. It possesses small, fleshy leaves and attractive, starry-white flowers. *Hylotelephium telephioides* (Allegheny Stonecrop) enjoys full sun and humus-rich sandy soil. It has glaucous leaves and pink starry flowers in flat-topped clusters. *Sedum glaucophyllum* (Cliff Stonecrop) prefers partial sun and well-drained gravelly or shale-based soil. It forms an extensive mat of compact rosettes of gray-green fleshy leaves resembling pinwheels. Numerous white four-petaled flowers sparkle above the foliage.

Sedum glaucophyllum

Native Irises add an attractive element to the rock garden. *Iris cristata* (Dwarf Crested Iris), found in partial shade and growing in rich, well-drained soil, reaches about six inches in height. Beautiful three-inch blue flowers with a golden crest nestle among the arching leaf blades. *Iris verna* (Dwarf Violet Iris) grows in sandy or rocky open woods. It has straight, narrow, bright green leaves that reach six inches in height. The showy, lavender flowers, however, do not possess the crest.

Pussytoes, whose name refers to the resemblance of the flower heads to cats' paws, make a welcome addition to the rock garden *Antennaria virginica* (Shale Barren Pussytoes) thrives in full sun to partial shade with poor shale-based or sandy soil. This low spreading plant forms attractive colonies of silvery rosettes with fuzzy white flowers. *Antennaria plantaginifolia* (Woman's Tobacco) can withstand very dry soils. The leaves are pale green above and woolly beneath. The flowers are white or purplish.

Phlox subulata (Moss Phlox) is densely matted with semi-evergreen, awl-shaped leaves. The flowers range from pink to rose-purple to white and possess a dark eye. It prefers dry, rocky habitats in full sun. *Phlox stolonifera* (Creeping Phlox), which prefers moist woods, has trailing stems that can root at the nodes. The flowers range from bluish to rose-purple.

Paronychia argyrocoma (Silverling) grows on open rocky slopes, outcrops, or ledges and forms mats or tufts. The numerous silky, hairy stems hide small white flowers that grow at the top of the stem.

Meehania cordata (Meehan's Mint) is a trailing plant that requires shade and grows in moist humus soil. Its heart-shaped leaves and pale blue, one-inch flowers will enhance any rock garden slope.

No rock garden should be without examples of native grasses and sedges. *Danthonia spicata* (Poverty Oatgrass) is a denizen of dry woods. Reaching only four inches in height, its foliage remains green during the winter months. *Luzula acuminata* (Hairy Woodrush) grows in partial shade with average soil. Its grass-like leaves are adorned with wispy, white hairs. *Luzula multiflora* (Common Woodrush) can withstand full sun. Its leaves are a beautiful reddish brown. *Carex glaucodea* (Blue Sedge) highlights a winter day with its blue foliage. *Carex platyphylla* (Broadleaf or Silver Sedge) likes dry, dappled shade. Its silvery green leaves stand out beside rocks partially buried in the garden.

The list of native plants suitable for rock gardens or hypertufa containers is extensive and varied. Joining your local chapter of the North American Rock Garden Society is another good way to increase your knowledge of native plants suitable for rock gardening.

HYPERTUFA and Native Plants 18

Native plants grow in every conceivable type of habitat. From the lowest meadow to the top of the mountains, native plants have adapted successfully to dry shale barrens, acidic bogs, swamps, prairies and deeply shaded woods. Most gardeners do not have this type of diversity in their landscape. Constructing hypertufa containers is one way to grow native plants from these diverse ecosystems.

Originally, these types of containers were made from tufa rock. Tufa is a soft and porous calcite rock that forms by precipitation from evaporating spring and river water. Because of its light weight and porosity, these containers were perfectly suited to grow alpine plants. Unfortunately, tufa became scarce and expensive due to excessive collecting.

Hypertufa is a handmade stone. It is simply a combination of cement, peat moss, and sand that can be molded to create garden containers. Originally developed as a way to grow difficult alpine plants from around the world by providing the unique growing conditions that these species required, hypertufa offered native plant gardeners a similar means to create the necessary environment of many temperamental native plants. Also, hypertufa containers allow gardeners without a lot of garden space to grow several native plants in a small area.

Constructing a hypertufa container is an entertaining way to spend a rainy afternoon. The first step is to select the mold in which you will shape your container. Discarded metal or rubber dishpans and hospital bedpans make excellent molds. The three basic ingredients are portland cement, mason's sand, and milled peat moss. To create a porous texture like tufa mix you will need one part portland cement, one part mason's sand, and two parts milled peat moss. You want your mix to be the consistency of bread dough—not too wet and not too dry.

To begin making your hypertufa container, cover your molds with a thin sheet of plastic. You can use plastic garbage bags or supermarket bags. Make sure the entire surface of the mold is covered in order to prevent the mixture from sticking to the sides. Begin filling the bottom of the mold to a depth of one and a half inches. Press the mixture firmly to eliminate any air holes or pockets. After creating a drain hole in the bottom, you can start building the walls making them two inches thick and being sure that you have packed the mixture tightly against the mold.

After you have achieved the desired shape, set the container aside and allow it to dry for forty-eight hours. When the container is dry enough to handle, gently remove it from the mold. Be sure that all of the plastic has been removed from the surface of the container. To give it a weathered appearance take a wire brush and

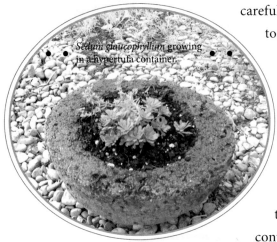

Sedum glaucophyllum growing in a hypertufa container.

carefully distress the surface to remove any sheen left from the plastic and also to make sure that any thin edges or protuberances are eliminated. You can also use a gouge or chisel to create the impression that the container was sculpted from solid stone.

Set the container aside in a shaded spot, and let it finish curing for five weeks. Before planting anything in the hypertufa container, you must thoroughly rinse it to remove any traces of the chemicals released in the cement.

The soil mixture for your hypertufa container should consist of screened leaf mold, peat moss, coarse sand, and finely shredded pine bark. To ensure good drainage, place a one-half inch layer of gravel in the bottom of the container. You should also use either turkey grit or small pebbles as mulch around your plants. This will help prevent the soil mixture from drying out.

What can you plant in your hypertufa container? One thing I have done with my smaller containers is to plant individual species, especially ones that are difficult to sustain in a garden situation. *Draba ramosissima* (Rock Twist) is a beautiful plant that grows on shale barrens. It is truly one of our alpine species from the Allegheny Mountains, and it is not well known among native plant enthusiasts. From a four-inch bun of tight foliage, it produces sprays of tiny white flowers that resemble Gypsy's Breath. Quite often, the seeds will fall into the soil mixture and produce seedlings the next year.

Native *Sedums* are also excellent choices. *Sedum ternatum*, with its dark green foliage and bright white flowers, will fill a container and droop nicely over its sides in a shaded spot. *Sedum glaucophyllum*, which has a lighter green foliage and cream-colored flowers, will perform similarly but can take a little more sun. And *Hylotelephium telephioides*, with its glaucous leaves and pink, starry flowers, will reach a foot in height.

If you construct a container that is a minimum of one foot by two foot in size, you can create a small habitat for several different species. One example would be to plant *Ionactus linarifolius* (Stiff Aster) in the center of the container. Its foliage resembles Rosemary, and its blue petals and yellow corona will brighten your garden in early fall. On one corner, you could plant *Campanula rotundifolia* (Bluebell Bellflower). Its delicate and dark purple flowers bloom in early summer. On the opposite corner, you can use *Antennaria virginica* (Shale Barren Pussytoes) with its silvery rosettes of fuzzy leaves. *Iris verna* (Dwarf Violet Iris) can be planted at another corner. Its purple flowers appear in early spring. And finally *Meehania cordata* (Meehan's Mint), with its one-inch pale blue flowers, could be placed at the last corner and allowed to spread over the sides of the container.

Growing native plants in hypertufa containers will become a new and fascinating gardening experience. You will be able to enjoy the beauty of plants that grow in diverse habitats and are not often seen by gardeners. Your choices of what species to grow is limited only by your enthusiasm and the size of your hypertufa container.

Xyris torta

GRASS-LIKE Plants 19

Native grasses, while some of the most attractive plants in the landscape, do not necessarily fit into the overall design of every gardener. Native plants with grass-like appearance, however, can make striking additions to your garden. Not only can they be substituted for grasses, they offer unusual inflorescences throughout the growing season. And, once they have finished blooming, they continue to contribute to the structure of the garden because their foliage resembles clumps or tufts of grass. Using these grass-like plants offers an added dimension and provides an opportunity to use species rarely seen in gardens.

Blue-eyed Grasses, members of the genus *Sisyrinchium*, are plants related to Irises, not grasses. The flowers are yellow-eyed with varying shades of blue petals and appear either in a small umbel or as a solitary bloom. All of the species have tufts of grass-like leaves and flattened winged stems that resemble the leaves. They grow in open woods or clearings and begin to bloom in spring and may continue to flower until fall. *Sisyrinchium angustifolium* (Narrowleaf Blue-eyed Grass) has a distinctly winged lower stem. The flowers are frequently pale blue. *Sisyrinchium atlanticum* (Eastern Blue-eyed Grass) is barely winged on the lower stem and has pale green foliage. The flowers are violet-blue. It is mostly a Coastal Plain species and is found in low wet places. *Sisyrinchium albidum* (White Blue-eyed Grass) has light green leaves that are mostly basal. The flowers are white or pale blue. It prefers open areas in dry woods. *Sisyrinchium mucronatum* (Needletip Blue-eyed Grass) is darker green and has very narrow leaves. Its pale blue flowers rest above a reddish purple spathe, a leafy bract that surrounds the inflorescence.

Common Goldstar (*Hypoxis hirsutus*) possesses similar grass-like leaves with six-petaled yellow flowers. Seldom reaching more than eight inches in height, it should be planted in groups; and in time it can form an attractive but small groundcover. It prefers dry open woods, but is highly adaptable.

Yellow-eyed Grasses are members of the genus *Xyris*. The three-petaled yellow flowers of *Xyris* species are ephemeral (short-lived). To fully appreciate their beauty, the plants should be placed in groups. *Xyris torta* (Slender Yellow-eyed Grass) grows to twenty inches and has spirally twisted and very narrow leaves. The base of the plant is enlarged, appearing as if an enlarged bulb. *Xyris caroliniana* (Carolina Yellow-eyed Grass) has somewhat broader leaves which are not twisted. The base of the plants is soft and flat. *Xyris* plants prefer wet areas.

Acorus americanus (Sweetflag) is a wetland species with a thick rhizome and basal sword-shaped leaves. This species should not be confused with *Acorus calamus*, which is a non-native species. One of the interesting characteristics of Sweetflag is the inflorescence which is a tapered spadix that is completely covered with small brownish or greenish flowers. It projects at an angle from the middle of the stem. Although an aquatic plant, it can be grown in moist garden soil. It can spread by rhizomes and should be contained in a sturdy pot buried in the ground. The plant can be divided every three years.

Xerophyllum asphodeloides (Eastern Turkeybeard) forms a thick, bristly clump of stiff and needle-like leaves. The white flowers appear in a dense raceme that can reach ten inches in length. The flowers appear in late spring and early summer. It grows in dry woods or in pine barrens. When in bloom, it rivals any ornamental grass.

Woodrushes are attractive plants with grass-like leaves that enjoy sunny to partially shaded areas. *Luzula acuminata* (Hairy Woodrush) is adorned with wispy, white hairs. It blooms in early spring. *Luzula multiflora* (Common Woodrush) possesses leaves that are an appealing reddish brown. Both species maintain their foliage during the winter months.

The genus *Eleocharis* (Spike Rush) contains several species that are both similar as well as variable. Denizens of wet areas, they can form mats or small colonies. They produce leafless stems and a solitary conical flower

Sisyrinchium angustifolium

cluster at the top of each stem. *Eleocharis tenuis* (Slender Spikerush) has tufts of very slender leaves. *Eleocharis obtusa* (Blunt Spikerush) has thicker stems and grows in clumps.

Three species of Iris can also be used effectively in place of grasses. Their attractive blooms brighten any garden in spring. For the remainder of the year, their erect blades resemble clumps of sedge. *Iris cristata* (Dwarf Crested Iris) grows in dry to slightly moist soil. It will spread to form a small groundcover, but it can be divided every three to five years. *Iris verna* (Dwarf Violet Iris) has narrower leaves and a beautiful purple bloom. It prefers well-drained soil and spreads at a much slower pace. *Iris lacustris* (Dwarf Lake Iris) prefers wet areas. It also has a purple blossom and will slowly create a small colony that can periodically be divided.

Equisetum hymenale (Scouringrush Horsetail) has evergreen hollow stems with no branches. It gives the appearance of a clump of bamboo. Toothed black bands add an attractive quality to the stems. Although it prefers wet places, it can be used in drier garden soil if placed in a buried container about the size of a five-gallon bucket and is periodically divided. It can also be grown in a large patio container. Equisetum is the single surviving genus of a class of primitive vascular plants that date back more than 350 million years. Scouring Rush derives its name from its stems being used by early settlers to scour pots and pans. Furniture makers used the rough stems like sandpaper to create a satiny finish to wood.

The character and structure of a native plant garden is only limited by the enthusiasm and imagination of the gardener. Using grass-like plants adds variety and interest and will please any visitor, whether they are fellow gardeners or curious insects.

CREATING a Native Prairie 20

Prairies are vast grasslands that once dominated the landscape of the heartland of North America. Early settlers, who had spent weeks winding their wagons through dense forests in a search of new lands to farm, were amazed to find these treeless expanses. For many of them, trees were indicators of fertile soil. They quickly bypassed these "barren" lands and searched for more habitable areas. For those who chose to settle in the prairies, farming was not practical because the deep sod formed by the extensive root systems of the grasses made it nearly impossible to turn the soil. Not until the development of the steel plow would the prairies succumb to agriculture.

Native prairies have almost completely disappeared from the American landscape. "The prairie flowers have strangely enough disappeared from open ground, under the croppings of cattle and the clippings of the scythe," observed one visitor to the Midwest in 1847. "Only a half dozen sorts were seen in a ride of 30 miles, and these straggling at great distances." Overgrazing, excessive plowing, and land development have destroyed this fragile ecosystem. Only small remnants of these prairies remain to give us a glimpse of one of the wonders of nature. These relic prairies, although small in size, can still be found in old settlers' cemeteries, along railroad rights-of-way, or in the corner of a farmer's field.

More than two hundred different plant species grew in the prairies of the Midwest. Asters, Sunflowers, Goldenrods, Coneflowers, and Blazing Stars created a rainbow of colors. And different species of grasses served as the protective blanket for this panorama. Some of these prairie grasses grew "taller than a man on horseback."

Efforts have been made to reestablish prairies. Seeds are collected from the relic prairies and planted in protected areas. The original ecosystem of the prairies, however, can never be recreated by man. We can, nevertheless, appreciate some of the beauty of these plants and the complexity of this unique ecosystem by creating a native prairie in our own gardens.

It is important to start on a small scale. A four-foot by eight-foot plot can accommodate several different species and will be easy to manage. The first step is to remove all of the existing vegetation by scalping

Indian Grass prairie in southeastern Pennsylvania.

the turf to a depth of two to three inches. Do not till the soil because this will only bring buried weed seeds to the surface. Dig a cavity large enough for each plant and amend it with compost. Leave the remaining soil undisturbed.

Prepare a map of the garden and make a selection of the prairie species to be planted. In the center of the bed, plant alternately *Andropogon gerardii* (Big Bluestem), *Silphium laciniatum* (Compass Plant), and *Silphium terebinthinaceum* (Prairie Dock). On either side of this center section, plant a combination of *Liatris aspera* (Tall Blazing Star), *Liatris pycnostachya* (Prairie Blazing Star), *Asclepias sullivantii* (Prairie Milkweed), *Oligoneuron ohiensis* (Ohio Goldenrod), *Eryngium yuccifolium* (Rattlesnake Master), and *Schizachyrium scoparium* (Little Bluestem). Along the border of the small prairie, plant *Boutelea curtipendula* (Side Oats Gamma Grass), *Hypoxis hirsutus* (Common Goldstar), *Gentianopsis crinita* (Greater Fringed Gentian), *Phlox glaberrima ssp. interior* (Smooth Phlox), *Coreopsis palmata* (Stiff Tickseed), *Amorpha canescens* (Leadplant), and *Viola pedata* (Birdfoot Violet).

These plants, once established, will attract butterflies, birds, and many beneficial insects. And you will have a small parcel of what was once a distinct plant community.

Monarch butterfly pollinating a Blazing Star.

THE Forgotten Pollinators — 21

As winter loosens its grip and we see glimmers of the spring to come, many gardeners get restless, hankering to get outside and start gardening. During my walks through the gardens in late winter and early spring, I look for any signs of new growth or damage to the plants. I am always amazed to see green buds on plants that only a week ago were covered by snow and ice. Then I remember that snow is Mother Nature's blanket.

As you look forward to the new year of gardening, think about all of the pollinators that make possible the flowers, seeds, and harvest which we so enjoy. Pollinators tend to be one of the gardening experiences we often overlook. In fact, many gardeners unaware of the importance of these pollinators spray pesticides and herbicides with reckless abandon, eliminating both beneficial and harmful insects and animals.

The time has come to begin gardening *for* pollinators. There are simple and inexpensive things you can do to increase the number and diversity of pollinators living on your land. You need to become familiar with the different habitats on your property that shelter and support pollinators. After you have identified where these pollinators are foraging and living, you need to take the necessary steps to protect them from disturbance and pesticides.

Because many home landscapes do not offer these natural habitats, each of us must provide new habitats by planting flowering plants and creating new nesting sites. Using native plants is essential to the successful introduction of these pollinators. Most native plants have specific insects or animals that can pollinate their flowers. Without these pollinators you may be able to enjoy the blossoms, but there will be no viable seeds at the end of the growing season. And without viable seeds, the continued presence of these species is lessened each year.

Plants must have insects to exist. Humans also must have insects to exist. Approximately 80 percent of the different plant species worldwide depend on pollination by animals (most of which are insects). Sadly, the population of wild pollinators is declining around the world, because of loss of habitat and the rampant use of pesticides and herbicides.

Vast numbers of insect species are needed to pollinate these plants. What we have only recently begun to understand is that a specific relationship has evolved between these plants and their special pollinators. The shapes and colors of each species of flowering plant, their unique scent, the location of the flower on the stalk, the time of the season and daily schedule of the availability of pollen and nectar are adapted exactly to attract particular species of insects.

When this unique relationship is disrupted or destroyed, both plants and their pollinators face extinction.

Native bees are one of our most important pollinators. Like all pollinators, native bees have critical requirements. Bees eat only pollen and nectar. While gathering these food resources, they move pollen from one flower to another and in the process, they pollinate the plants. Native bees depend on an abundance and variety of flowers. Furthermore, they must have access to flowering plants throughout the growing season. Creating gardens with native plants is essential to the survival of native bees.

Native bees, unlike honeybees or wasps, do not build wax or paper structures. Nevertheless, they still require places to nest. Wood-nesting bees tend to be solitary, and frequently build nests in the twigs or beetle tunnels found in dead trees. Ground-nesting bees construct their nests in tunnels under bare earth. And cavity-nesting species, such as the bumblebee, occupy small spaces such as abandoned rodent burrows.

To ensure the survival of all pollinators, use no pesticides in your gardens. Almost all insecticides are deadly to native bees. Equally important, the use of herbicides kills many of the flowering plants necessary to the survival of all pollinators.

What can you do? Keep tillage of your gardens to a bare minimum. Many of the nests of pollinators are underground. When the soil has been overturned, you have inadvertently destroyed these nests. Provide a wide range of native plants to satisfy the food requirement of pollinators throughout the growing season. One way to ensure a successful harvest from your vegetable garden is to interplant native wildflowers. Their blooms will attract a wide range of pollinators who will eat harmful insects and pollinate the flowers of your plants, ensuring that fruit will form. You can also create marginal areas near the gardens, planting hedgerows and windbreaks with a variety of native flowering plants that will attract pollinators and provide them with shelter.

Creating a native plant garden may not conserve entire populations or species of threatened or endangered pollinators and plants. However, undertaking such an endeavor can remind us of the importance of butterflies, hummingbirds, bees, ants, wasps and many other insects. "One butterfly or one wildflower does not an ecosystem make," cautions one naturalist. "Nature's rich complement of butterflies and flowers and other organisms and their myriad, evolved relationships can only survive if the diversity of the natural habitat is preserved." It is time to take a step in that direction.

CREATING a Rain Garden — 22

In recent years, the amount of damage to private and public property because of flooding has totaled billions of dollars. As mountains are reduced to negligible semblances of their former stature; as more highways are built and paved; as more and more new homes are constructed; and as more parking lots cover once productive meadows, the ability of the soil to absorb rainfall is limited. These impervious surfaces are areas that quickly shed rainwater into already overtaxed storm drains and nearby streams and rivers. The increased storm water runoff adversely affects every fiber of our natural environment and economy. Because this runoff is untreated, the amount of pollution from our own yards and gardens—which includes fertilizers, pesticides, herbicides, pet wastes, grass clippings, and yard debris—has increased exponentially.

One way to help prevent these pollutants from pouring into our precious waterways and contaminating underground water supplies is to create a rain garden. A rain garden, quite simply, is an attractive and landscaped area that has been planted with native wildflowers, grasses, and sedges that grow naturally in wetlands. These beautiful gardens are built in depressions that have been designed to capture and filter storm water runoff from rooftops and driveways around the home. Rain gardens offer countless benefits to the homeowner and the environment. A rain garden filters storm water runoff before it reaches local waterways. By slowing this runoff, it helps to alleviate flooding. Because it is actually a garden, it also enhances the landscape of individual yards. These native plants in these rain gardens create a habitat and food source for wildlife, especially birds, butterflies, and bees. And the rain garden can safely recharge the ground water supply. Constructing a rain garden does not have to be an expensive or complicated

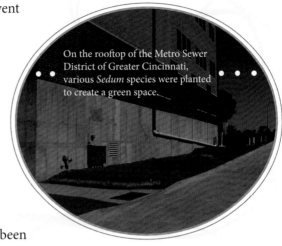

On the rooftop of the Metro Sewer District of Greater Cincinnati, various *Sedum* species were planted to create a green space.

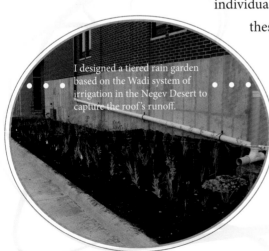

I designed a tiered rain garden based on the Wadi system of irrigation in the Negev Desert to capture the roof's runoff.

endeavor. A number of factors—the size of your yard, the amount of money you wish to invest, and the type of garden you want to create—will determine the design of your rain garden. Whether your rain garden is large or small, it will help to solve water quality and flooding in your area *and* it will be a welcome and satisfying addition to your landscape.

Where should you build your rain garden and how large should it be? Your first task is to determine the drainage pattern of your property. Where does runoff flow? Are there areas where rainwater collects? Rooftops, paved surfaces, slopes, and compacted soils are usually the largest sources of runoff. When you have determined this information, you can find the proper location for your rain garden. A rain garden should be a minimum of ten feet from your house and your neighbor's home to prevent any damage from water seepage. Never place a rain garden over or close to the drain field of a septic system. *Do not* place your rain garden in the area where water currently collects. Instead, it should be placed up the slope from such areas in order to reduce the amount of water flowing into these wet areas. Rain gardens do best in a sunny or at least partly sunny location. Integrate your rain garden into your landscape with either a formal or informal design. Never place a rain garden beneath a large tree, because it might damage the root system of the tree. *Be aware of underground utilities!*

Once you have decided where to build the rain garden and the size it will be, you can begin to excavate and level the site. If the area has an existing lawn, you will have to remove the turf. Although herbicides can be used, this would be contrary to our purpose of eliminating pollutants. I recommend either placing black plastic over the turf to block sunlight or scalping the grass with a mattock. (You can also rent a sod cutter for

Bright pink flower of *Chelone lyonii*.

larger areas.) Determine the perimeter of the rain garden. You can use a garden hose, rope, or stakes to define the area to be excavated. If your lawn is flat, dig to a uniform depth, about four inches, throughout the garden. If the lawn has a slope, you will have to dig out the high end of the garden. The depth of a rain garden on a slope should not exceed twelve inches. The excavated soil can be used to raise the low area. Use the excavated soil to create a berm on the downhill side of the garden. The berm is a low earthen mound that surrounds three sides of the rain garden and will help to hold water during a storm.

You are now ready to plant your rain garden. This is an opportunity to enhance the beauty of your landscape. Choose native plants that are tolerant of both dry and wet conditions. By selecting a diversity of species you will not only create a stunning visual effect but also have healthier plants. Another factor to consider is the height, the bloom time, and the color of each plant. By using plants with different blooming times, you will extend the flowering season of the rain garden. Using different heights and shapes will provide depth and texture. And randomly planting individual species in groups of at least three, repeating this throughout the garden, will make a bold statement of color. Finally, incorporate not only wildflowers but grasses, sedges, and rushes to create a thick underground root system that will absorb the water.

Rain garden planted at the Exhibition Coal Mine in Beckley, West Virginia.

Using native plants is the key to success with your rain garden. Native species are adapted to your local area. They do not require fertilizers. They do not need pesticides to thrive. And they survive downpours and droughts. The list of native plants to use in a rain garden is vast; following are a few suggestions.

Chelone glabra (White Turtlehead) offers spikes of elegant turtlehead white flowers, while *Chelone lyonii* (Pink Turtlehead) produces brilliant rose pink turtlehead flowers. The Baltimore Checkerspot Butterfly will find your rain garden a favorite breeding site. *Filipendula rubra* (Queen of the Prairie), with its showy pink flowers, will sway to and fro in the breeze. *Liatris spicata* (Dense Blazing Star) sends up dense spikes of purple flowers. *Lobelia siphilitica* (Great Blue Lobelia) and *Lobelia cardinalis* (Cardinal Flower) will brighten the edges of your rain garden with deep blue and red flowers. *Physostegia virginiana* (Obedient Plant), with its purple to pink flowers, should be planted along the inside edge of the berm. Its upright stems and dense roots will stabilize the soil of the berm. *Eupatoriadelphus dubius* (Coastal Plain Joe Pye Weed) has rounded pink flower heads and will provide stature without overpowering the rest of the garden. *Iris versicolor* (Harlequin Blue Flag), *Iris virginica var. shrevei* (Shreve's Iris), and *Iris fulva* (Copper Iris) lend the appearance of cattails while offering colorful blooms. *Symphyotrichum novae-angliae* (New England Aster), with purple to pink flowers, and *Eurybia radula* (Low Rough Aster), with lavender flowers, brighten the garden in late summer and early fall. *Asclepias incarnata* (Swamp Milkweed) will attract countless butterflies with its pink blossoms.

And do not forget the grasses and grass-like plants. *Andropogon gerardii* (Big Bluestem) and *Spartina pectinata* (Prairie Cordgrass) will tower above the wildflowers and produce interesting inflorescences.

Elymus virginicus (Virginia Wild Rye) can be planted along the garden's edge. *Glyceria striata* (Fowl Mannagrass) has unusual flower heads. *Juncus torreyi* (Torrey's Rush) has round, orange seed heads. And *Carex grayii* (Gray's Sedge) has seed heads that resemble a medieval mace.

Many other native species will grow successfully in your rain garden. The selection is limited only by your imagination. The reward will be a garden of native plants that becomes a home to countless beneficial insects and other wildlife and also helps to protect our precious waterways.

Eriogonum allenii growing in shale barrens near Covington, Virginia.

THE SHALE Barrens of Central Appalachia 23

Shale barrens are one of the most geologically interesting and botanically intriguing areas in West Virginia and Virginia. Comprised of sparse woodland, shrub land, and open herbaceous rock outcrops, they occur on ridge and valley shale of the central Appalachian Mountains. These botanic communities are endemic to western Virginia, West Virginia, west-central Maryland, and south-central Pennsylvania. In most cases, plants are found growing on steep, eroding slopes of thinly bedded and weathered shale which face south to west. They have sparse tree cover and very little soil.

Nevertheless, these shale barrens are home to some of the rarest plants growing in the Appalachian Mountains. Most of these species are endemic to the shale barrens and are commonly referred to as shale barren endemics. In 1911, Edward S. Steele coined the term "shale barren" to describe these unique plant habitats. Referring to a shale barren in Virginia, Steele characterized it as "one of the most fascinating spots in which it has been my fortune to botanize." He was not the only person who thought so. Several noted botanists and scientists of the late nineteenth century vacationed at the world-renowned White Sulphur Springs resort. During their forays into the nearby mountains, they discovered the classic shale barrens located on Kate's Mountain. They found several species of plants that were new to science. Specimens were collected and sent to various herbariums, especially the New York Botanical Garden which had been established in 1896. White Sulphur Springs became the type location for at least eight of the shale barren endemics.

Visiting the shale barrens is a unique experience. They are hot, dry, and unstable. There is the occasional copperhead and on rare occasions an encounter with a black bear. But a visit to the shale barrens is worth every step just to see the rare and endemic plants and animals and the extreme conditions in which they grow. A combination of geology, soil, topography, and climate created the shale barrens. Beneath it all is bedrock: shale. Because shale is a highly friable rock, small fragments—called channers—tumble down the slopes and create a very unstable substrate. There is very little soil. And water sheds easily, making these barrens extremely dry and inhospitable. These unusual habitats, however, support rare species.

The most commonly found trees are extremely scrubby forms of Chestnut Oak, Virginia Pine, Eastern Red Cedar, and Pignut Hickory. Growing along the fringe of the barrens are white ash, table-mountain pine and shagbark hickory. A few shrubs also grow on the barrens, including Shadbush, Black Huckleberry, Dwarf Hackberry, Deerberry and Bear Oak.

Despite these extreme conditions, there are some animals that reside in the shale barrens. Among the reptiles are five-lined skinks, eastern fence lizards, wood turtles, copperheads, and timber rattlesnakes. Bird watchers will find pine warblers, prairie warblers, Carolina wrens, and broadwing hawks. Several species of skippers, butterflies and moths also frequent the shale barrens. Mammals, both large and small, can be found on the shale barrens. White-tailed deer, eastern gray squirrels, foxes, coyotes, black bears, bobcats, and eastern red bats frequent these barrens.

It is the rare plant communities, however, that have drawn attention to these shale barrens for over a century. Kate's Mountain Clover (*Trifolium virginicum*) first drew attention to the uniqueness of the flora of the shale barrens. During an exploring trip in 1892, Dr. John Kunkel Small discovered *Trifolium virginicum*. His discovery prompted other botanists to visit this unique area, and before long several other species were identified.

Phlox buckleyi (Swordleaf Phlox) was found by Samuel B. Buckley in 1838 at White Sulphur Springs, but remained unnoticed in his herbarium for eighty years. Edward T. Wherry formally named the plant in 1930, a century after its discovery. Although *Phlox buckleyi* does not grow on the barrens, it is usually found near the bases of these shale slopes. It has unbranched linear leaves with vivid pink flowers. *Clematis albicoma* (Whitehair Leather Flower) was first collected by Gustav Guttenberg in 1877 on Kate's Mountain. This is a rather short species with dull purple and solitary flowers that grow at the end of the branches. *Eriogonum alleni* (Shale Barren Buckwheat) is named for its discoverer Timothy F. Allen. This genus is typical of those found growing in the Rocky Mountains and was thought at one time to have been introduced by dust storms carrying the seeds. It prefers the barest and most sterile growing conditions. The dark green leaves have loose tufts of soft white hairs. The flowers are dark yellow and grow in flat-topped clusters. *Oenothera argillicola* (Shale Barren Evening Primrose) was discovered by Kenneth K. MacKenzie, an amateur botanist, in 1903 at White Sulphur Springs. When in bloom, this is one of the more spectacular sights on the shale barrens. On a recent trip to the shale barrens, I saw a shale slope that appeared as if an artist had brushed it a bright yellow. *Taenidia montana* (Mountain Pimpernel) was also collected by MacKenzie in 1903 on Kate's Mountain. This rare, monotypic species is restricted to the Allegheny Mountains. It has thickened roots and thin umbel rays of yellow flowers. *Packera antennariifolius* (Shale Barren Ragwort) was collected in 1897 by Nathaniel Britton on Kate's Mountain. The tufts of white wooly leaves are quite distinct. The erect stems support showy white flowers.

Arabis serotina (Shale Barren Rockcress) has become one of the most endangered species. Shale Barren Rockcress is an erect biennial and a member of the mustard family. It produces a white inflorescence. Because it is one of the most restricted shale barren endemics and is known from only sixty populations with fewer than one thousand individuals, the threats to this species apply equally to all of the shale barren endemics. Spraying insecticides to control gypsy moth has had a devastating effect on the pollinators of *Arabis serotina*. Road construction, railroad construction, hiking trails, and dam construction have reduced or destroyed its habitat. Several invasive species, including *Centaurea maculata* (Purple Knapweed), can out-compete *Arabis serotina*. Deer browsing, long periods of drought, and the naturally low populations of all shale barren endemics present potential threats.

Shale barrens are one of our botanical treasures. They must be protected so that future generations will have an opportunity to see and enjoy the unique flora and fauna that call these habitats "home."

UNSUNG Heroines in the Movement to Preserve Native Plants

24

During the nineteenth century, it was extremely difficult for women to enter science as professionals. Botany, however, opened its doors to a select few whose talents were too significant to go unnoticed. Scientific illustration was one of the acceptable ways for women to enter the professional ranks in the nineteenth century. In a male-dominated society, art was deemed an appropriate talent for women to develop and there were many gifted artists who made significant contributions to the growing knowledge and understanding of the native flora of North America. There were others whose pioneering spirit took them into unexplored territories and brought to light several native plants new to science. And there were others who sought to preserve the native flora before it was lost to urban and economic development.

Lilla Leach was one such pioneering botanist. Lilla was born on March 13, 1886 in Barlow, Oregon. From childhood, Lilla exhibited a keen interest in wildflowers. Lilla's father, William Irvin, sent her to Tualatin Academy, a preparatory school, in 1904. It was there that she met her future husband, John R. Leach. Together, they set out to find the botanical treasures of the Siskiyou Mountains in southwestern Oregon from 1928 to 1938. Because there were no roads, and quite often no trails, Lilla and John used burros to pack their gear on these long treks. Describing his camp duties, John said, "I am the muleskinner...acting as a buffer between the botanist and the burros." Their travels were filled with adventure, encountering cougars, bears, and rattlesnakes. There were also humorous times, such as the day the burros ate John and Lilla's underwear while they were swimming in the Chetco River.

Lilla graduated from the University of Oregon in 1908 with a degree in science and with an emphasis on botany. She taught science classes in the Eugene public school system for three years after graduation. Because it was customary at that time for women teachers to end their careers after they married, Lilla filled in as a clerk at her husband's pharmacy. John and Lilla were soon collecting native plants close to highways and back roads. It was not long before Lilla decided it

Lilla Leach collecting wildflowers in the Siskiyou Mountains.

was time to botanize in unexplored areas. In 1930, Lilla made the first scientific collection of *Kalmiopsis leachiana*, a relict plant from the Tertiary geologic period. During their ten-year sojourn, they discovered five plant species new to science. Their plant collections became world renowned. Their botanical efforts resulted in the creation of the Kalmiopsis Wild Area which encompasses over 70,000 acres.

Emily Hitchcock Terry (1838-1921) was born in Amherst, Massachusetts. Her parents were highly intellectual and artistic. Emily's ambition was to become an artist, but she also had a deep interest in botany. In time, she merged her interests in art and botany and produced a collection of forty-nine watercolor paintings of the flora of Minnesota. Most of the paintings were of wildflowers. These paintings are part of a large collection, *American Flowers*, which contained 142 paintings from various parts of the United States. Unlike many plant collectors, Emily, instead of accumulating a typical herbarium of pressed specimens, created a "painted herbarium." Her paintings are the earliest known botanical illustrations of Minnesota. "As long as I live I shall work in botany," Emily remarked, "if I have any eyes to see."

Eloise Butler (1851-1933) was truly a Victorian plant hunter. She was another aspiring botanist who began her work at a time when science and career did not mix well with marriage. Eloise chose science and became a teacher in Minneapolis. Minneapolis was a small but rapidly growing mill town in the late nineteenth century. Natural areas filled with wildflowers that Eloise had come to know through her botanizing but began to disappear at an alarming rate as city streets and buildings were constructed. Alarmed by this rampant loss of rare and common species, Eloise helped to establish the Wild Botanic Garden, which soon was enveloped by the growing city. She saved countless plants from wanton destruction and moved them to new homes in the garden. The Eloise Butler Wildflower Garden and Bird Sanctuary, as it is now known, is the oldest existing wildflower garden in the United States. Eloise's pioneering work surely influenced the creation of other wildflower sanctuaries, but her significant contributions to the native plant preservation movement have gone unheralded unless, of course, you have been among the fortunate visitors to this amazing garden and have walked among plants once common but now so rare that if not for her efforts they would have been lost to future generations.

Eloise Butler

Marianne North (1830-1890) was an unmarried Victorian lady who came from a wealthy family. In 1871, she began a series of botanical expeditions to paint the tropical and exotic plants of the world. By the end of her journeys, she had made a collection of more than eight hundred paintings that are now housed in the Marianne North Gallery at Kew in England. Marianne used her paintbrush as the modern botanist uses a camera. Several of the plants she painted were almost unknown to botanists and horticulturalists. In fact, four species were unknown to science and were named in her honor. Many of the countries Marianne visited still retained areas that had not been spoiled by man. In California, however, she noted the "regrettable ravages" inflicted on the Redwoods. "It broke one's heart to think of man, the civiliser, wasting treasures in a few years to which savages and animals had done no harm for centuries."

Marianne was anxious to visit the United States because she felt that many people in England did not realize the vast botanical differences between North and South America. In 1871, she arrived in Boston. Her letters of introduction gave her access to many prominent politicians and scientists who went out of their way to provide assistance. As she traveled throughout New England, Marianne painted vivid landscapes of the rocky coasts, the autumn colors of the White Mountains, and the magnificent falls at Niagara. Her paintings of wildflowers captured the beauty and diversity of the flora of New England. One painting, in particular, is virtually a collage of Jack-in-the-Pulpit, Large Yellow Lady's Slipper Orchid, Solomon's Seal, False

• Marianne North at her home in Ceylon. •

Spikenard, and Wild Geranium. Another painting of wildflowers in the neighborhood of New York brings to life Pink Lady's Slipper Orchid, Naked Broomrape, Wild Columbine, and Pinxter Flower. Nearly a decade later, Marianne returned to the United States arriving this time in San Francisco. Her paintings of the trees, shrubs, and wildflowers of California and the Southwest illustrate the striking colors of these unique landscapes.

In 1879, she wrote to Sir Joseph Hooker offering to assume the financial cost to build a gallery to house her collection of paintings, to which he readily agreed. For the first time, the everyday citizens of England could see the beauty and diversity of the flora of North America in contrast to other exotic lands.

Without the pioneering efforts of these unsung heroines (and many others like them), the movement to preserve our native flora would surely have been delayed. It is hard to imagine how many more species would have been lost to future generations.

E. LUCY BRAUN and the Relic Prairies of Adams County, Ohio

25

In the summer of 1838, Dr. John Locke, assistant state geologist, made a reconnaissance survey of Adams County. Ambling about the county in a one-horse wagon that carried his equipment, Locke set out to see a unique rock formation known to the local inhabitants. Accompanied by a local guide, Locke headed east out of West Union, and turned south towards the Ohio River. Before long, Locke saw the dark column of dolomite above him on a ridge to his left. After climbing the steep hill to get a closer view, he noticed a fringe of grasses and wild flowers growing in the thin soil along the outer edge on top of the dolomite. Locke recognized these plants as being typical prairie species. He described in detail the rock formation and compiled a list of the plants. Buzzardroost Rock, as this geological feature came to be known, would continue to draw attention from both botanists and geologists.

It was not until nearly a century later, however, that Dr. E. Lucy Braun, accompanied by her sister, Annette Braun, made a visit to Adams County. Immediately impressed by the area, they soon began serious research on the unusual mixture of forests and prairies along the edge of Appalachia. Braun was born in 1889 in Cincinnati, where she lived for the remainder of her life. Lucy and Annette attended the University of Cincinnati. Lucy studied botany and geology and earned a Ph.D. in botany, becoming the second woman to do so. Annette was the first, receiving her degree in entomology. Beginning in her early twenties, and often accompanied by her sister, Lucy would take the Norfolk and Western Railroad to either Peebles or Mineral Spring, Ohio where they would be met by a local farmer with a horse and buggy. As Lucy became more and more familiar with these remnant prairies, she realized that they had to be protected from destruction by loggers and farmers. She worked tirelessly and, in 1959, the Cincinnati Garden clubs, along with the Nature Conservancy, donated funds for the initial purchase of forty two acres of Lynx Prairie, which has now grown to over three thousand acres.

As her research on deciduous forests progressed, Dr. Braun developed the theory that the southern Appalachians were the center of survival during the Pleistocene Period and dispersal of forest communities. She firmly believed that the

E. Lucy Braun

deciduous forest zone survived on the Appalachian plateaus in southern Ohio and Kentucky when the glaciers extended southward into Ohio. During these glacial epochs, some species presently having a spotty distribution near the glacial boundary may have persisted in favorable habitats. The northern extent of many southern and Appalachian species ended abruptly at varying distances south of the glacial boundary.

More than twelve hundred species of vascular plants are found in the 115 small remnant prairies that comprise what is known today as the Edge of Appalachia Preserve System. Five of these prairies are designated as rare throughout the world. Approximately one hundred of these plants are included on Ohio's roster of rare and endangered species. Perhaps the rarest species is *Paxistima canbyi*, known as Mountain Lover, Cliff Green, or Rat-stripper. This plant is only known from two locations in Ohio. Research indicated that the colony growing in Adams County originated from a single plant. Unable to produce seeds, Canby's Mountain Lover plants are clones reproduced by rhizomes sending up new shoots. Because these plants are genetically identical and cannot produce seeds, the colony exists in a precarious and fragile situation. Some believe that it was the search for this plant that brought Dr. Braun to Adams County in the first place.

More recent botanical explorations of these preserves have resulted in the discovery of species new to Ohio and also new to science. In 1988, a naturalist working for the Ohio chapter of the Nature Conservancy found a few individuals of a purple-flowered plant known as Ear-leaved Foxglove (*Agalinis auriculata*). This is a globally rare species. Another unique find was Juniper Sedge (*Carex juniperorum*) that was previously unknown to science. There will certainly be other new discoveries as botanists and other investigators roam the diverse habitats that make up these preserves.

Pellaea atropurpurea growing in a pocket of a dolomite boulder at Lynx Prairie.

Another fascinating aspect about the history of this region was the existence of the Teays River, which is now extinct. Not only were the glaciers responsible for the introduction of species into southwestern Ohio, they also changed the entire drainage system of this region. An ancient river known as the Teays originated in North Carolina and flowed northward, following the courses of the New River and Kanawha River, crossing Virginia, West Virginia, Ohio, Indiana, and Illinois before emptying into an area where the Mississippi River currently flows. The early glaciers created an ice dam blocking the flow of the Teays River. The water had no place to go and, as a result began to rise, ultimately creating a lake that stretched back into West Virginia and Kentucky. In time, the water reached to the top of the surrounding ridges and eventually overflowed them. One of the new streams formed was Ohio Brush Creek (its valley is home to many of these preserves), and the water from these streams emptied into the newly formed Ohio River.

The scattered prairie openings in Adams County are like islands in a sea of forest. My first visit to Lynx Prairie is etched in my memory. Walking through the cemetery leads you to the trail head and marker proclaiming the Lynx Prairie Preserve and National Natural Landmark. You enter the woods which are dominated by Virginia Pine and Red Cedar. Suddenly, the woods open up and you step into an open space filled with prairie grasses and wildflowers. Among the many species of grasses are *Andropogon gerardii* (Big Bluestem), *Schizachyrium scoparius* (Little Bluestem), and *Sorghastrum nutans* (Indian Grass). Scattered among these grasses are colonies of *Silphium terebinthinaceum* (Prairie Dock), *Matelea obliqua* (Climbing Milkweed), *Liatris scariosa* (Devil's Bite), *Liatris aspera* (Tall Blazing Star), and *Liatris cylindracea* (Ontario Blazing Star). As you continue along the trails, you will encounter *Asclepias viridiflora* (Green Milkweed), *Helianthus occidentalis* (Western Sunflower), and *Pellaea atropurpurea* (Purple-stemmed Cliffbrake). In fact, each and every step along the many trails in these preserves will lead you to a new discovery. It is not hard to imagine the excitement of Dr. Braun as she blazed these now well-worn paths.

E. Lucy Braun's unwavering efforts to protect these fragile and relic plant communities have been carried on by a legion of volunteers, scientists, and organizations. We all must do our part to preserve these ecosystems. They are the last refuges of rare and endangered species that have weathered the advance and retreat of glaciers and the wanton destruction of humans.

STEWARDSHIP: Who Will Take Responsibility? 26

During the 2007 Mountain State Arts and Crafts Festival held at Cedar Lakes, West Virginia, I had the opportunity to meet Darryl McGraw, the Attorney General of West Virginia. He and his staff were inviting folks to have their picture taken with him. When his entourage entered my tent, I was asked if I wanted my picture taken. Not realizing what was transpiring (I thought it was publicity for the festival), I gave them permission to take the picture. There was an older gentleman standing behind me. I asked him if he wanted to be in *my* photograph and put my arm around his shoulder. After the photograph was taken, his staff, somewhat embarrassed, introduced him to me. I thanked him for being in my picture. In turn, he asked me, "What kinds of plants are rare and endangered?" I responded, "At the rate the state is allowing the destruction of natural areas, all of them are potentially endangered, and the day will come when many of them will only be known in books or herbariums." By this time, a large crowd had gathered around the tent intent on hearing the conversation. McGraw asked me what I would propose as a solution to this problem. I quickly offered five ways to address the issue of invasive species and endangered native plants and agreed to prepare a report within two weeks for the Attorney General, on the condition that he would promise to follow through with the recommendations of the report. We shook hands and parted.

Following is the report that was submitted to West Virginia's Attorney General, and after it is a summary of what actually was done by the Attorney General's office.

Preserving and Conserving West Virginia's Native Plants

Economic development is essential to the welfare and growth of the state of West Virginia. This development, however, does not have to come at the expense of the sustainable natural resources that make West Virginia what it is: the Mountain State. In the past, special interest groups and private companies have ravaged the landscape through widespread timber operations, coal mining, construction of chemical and power plants, highway construction, new homes, and shopping malls. All of these economic ventures have come at the expense of the native flora and fauna of West Virginia. Each year, the number of threatened or endangered species has grown at an alarming rate.

Sadly, the loss of these plant and animal communities does not often reach the public's attention until long after the destruction has occurred. In fairness, what would West Virginia be without its magnificent mountains, rivers, and streams? The answer will become evident all too soon. What can be done to change how the natural resources of the state can be protected and conserved for future generations to enjoy? The answer lies in educating the public and state officials about how diverse and fragile these resources are and developing active programs at all levels to protect them.

I outlined five ways to begin this process. First, I urged the state to remove the sign on the bridges that claims West Virginia is "Open for Business." What this really means is that the resources are "For Sale." Many of the natural resources are finite. Once they have been exhausted, they cannot be replaced. Other resources may be replenished, but not necessarily in our lifetimes. The state needs to tout these diverse and natural resources as tourist attractions. West Virginia has far more to offer than just ski resorts and white water rafting. More and more of these natural resource areas must be set aside and protected. Once this is accomplished, the Department of Natural Resources, as well as private interest groups, could offer botanical excursions to tourists and citizens of West Virginia. The "Rails to Trails" program is an excellent example of this process. A more appropriate sign on the bridges could read: *Welcome to West Virginia: Pride of the Nation*; or the slogan so many citizens miss: *Wild and Wonderful West Virginia*.

Second, I suggested that the areas along the interstate highways that have signs indicating "West Virginia Wildflowers" actually grow native plants of West Virginia. Instead, those responsible for these areas have been planting primarily Cosmos and Dames Rocket, neither of which is a native plant of West Virginia. And even more disturbing is the fact that Dames Rocket is an invasive species that is now escaping into the state forests and other areas. These roadside areas are a wonderful opportunity to display the wealth of native plants in West Virginia. The same effort and expense to prepare and plant these areas could be directed towards planting native perennial wildflowers and grasses. Once established, they would provide a food source for birds and other wildlife. Afterwards, they would only have to be mowed once in the fall to cut down dry foliage. Each succeeding spring, these perennials would return more robust and floriferous than the previous year. And equally important, these areas would make visible to thousands of travelers the wealth and diversity of West Virginia's native plants.

Third, all along our interstates are rest stops for weary travelers. Within these areas, I recommended that small gardens be planted with a diverse selection of native plants. Each plant would have professional signage providing the botanical and common name of the plant along with a brief description of its flower and foliage. One such garden was planted at the Greenbriar exit off I-64 a few years ago and is maintained by the local garden club. I believe that if these gardens were created and planted with native plants they could be maintained by a combination of groups, including local garden clubs, members of the Master Gardeners program sponsored by the Extension Office at West Virginia University, the West Virginia Native Plant Society, and any interested people from the nearby communities. Brochures could be created that would provide additional information about the plants and areas and sites in the state that have unique native plants. Dolly Sods and Cranberry Glades are just two of the numerous locales within the state. There are many more and others yet to be identified and protected.

Fourth, the intentional and accidental introduction of alien species of plants is one of the dire threats to the natural resources of West Virginia. One need only drive along the West Virginia Turnpike and look at the rapid spread of Kudzu vines, *Paulownia* trees, and the ornamental grass *Miscanthus sinensis* to see how devastating these species can be. Invasive species are even more prevalent within the confines of our federal and state forests and parks. As mentioned before, *Microstegia vimineum* (Stiltgrass) is literally choking out hundreds of species of wildflowers and grasses. It is an annual grass that produces thousands of seeds per plant that attach themselves to any object that passes through them. ATVs are one of the main culprits. As they ride along trails covered with Stiltgrass, their tires spread the seeds wherever the ATVs venture. And all too often, the riders stray off the trails and traverse the sides of mountains or along gullies and cuts dissecting the slopes. Within a matter of weeks, there are green strips present where the tires have dispersed the seeds. Within three years it can replace all of the native vegetation on the floor of the forest. Tree of Heaven (*Ailanthus altissima*) is not only invasive but also poisonous. There have been instances where individuals who were sawing these trees became seriously ill from the sap. Another pernicious invasive plant that is a public health hazard is *Heracleum mantegazzianum* (Giant Hogweed). Originally from Asia and introduced as an ornamental plant, Giant Hogweed's clear, watery sap has toxins that cause photo dermatitis. Skin contact, followed by exposure to sunlight, produces painful, burning blisters that can develop into purplish or blackened scars.

There are far too many other species that have invaded the forests, meadows, and waterways of West Virginia. A statewide and concerted endeavor must be started to prevent and control the continued spread and introduction of these non-native species into the state. Efforts are already underway by both federal and state agencies to eradicate specific invasive species. But it will prove to be a fruitless battle if these same non-native species are allowed to grow in adjacent private lands and continue to be a source of seeds that will ultimately spread back onto state land. The general public, as well as state agencies, must be made aware of the ecological catastrophe that is taking place throughout the state because of non-native invasive plants.

Fifth, I urged the Department of Natural Resources to convene an annual conference to be held at different locations throughout the state devoted to the native plants of West Virginia and to the public health hazards and ecological destruction caused by invasive species. Speakers should represent a broad spectrum of the general public: professional scientists, state agencies (DNR, DOA, EPA), and private citizens who have invested many years in this field of study. These conferences should highlight the wealth of the state's resources, the myriad of problems facing these natural resources, and potential solutions to these problems. From these conferences would emerge viable strategies to preserve and conserve the native plants of West Virginia.

Summary of Attorney General's Office Investigation

The Attorney General sent copies of my report to the following officials in state agencies: Frank Jezioro, Director of the Division of Natural Resources; Gus R. Douglas, Commissioner of the West Virginia Department of Agriculture; John Pat Fanning, Chairman of the Senate Natural Resources Committee; William Stemple, Chairman of the House of Delegates Agriculture Committee; Joe Talbott, Chairman of the House of Delegates Natural Resources Committee; Earl Ray Tomblin, President of the Senate; Larry J. Edgell, Chairman of the Senate Agriculture Committee; Richard Thompson, Speaker of the House of Delegates; and Paul A. Mattox, Jr., Cabinet Secretary West Virginia Department of Transportation. Only two agencies provided a substantive response, and both are indicative of the attitude and stance of these agencies with regard to invasive species and endangered native plants.

With regard to wildflower plantings along the interstates in West Virginia, Paul A. Mattox stated, "Operation Wildflower, a joint program of the West Virginia Division of Natural Resources, the Garden Clubs, and the DOH began in 1990. One part of this program permitted individuals or organizations to sponsor sites including a sign that identified the sponsor or memorialized a loved one.

An attempt was made to plant native wildflowers but several problems were encountered. Among them was a difference of opinion on which species were native, seeds were impossible to find or were prohibitively expensive, and the native species did not do well in roadside growing conditions. In addition, the sponsors expected showy, colorful plots the first year which required annual species and ruled out perennial species which do not bloom the first year. All of the factors have combined to encourage the program to plant the annual species being used."

Commissioner of Agriculture, Gus R. Douglas, added the following comments:

"A number of my staff has also commented on the wildflower plantings along our highways and the use of the term 'West Virginia wildflowers.' In regard to these wildflower plantings there is really little the West Virginia Department of Agriculture…can do to change what is happening."

The West Virginia Department of Transportation's Division of Highways operates the Operation Wildflower Program and they are the agency responsible for what is happening. According to the DOH website, one of the objectives of this program is to "encourage the preservation of natural stands of native wildflowers that traditionally had been mowed." Another objective of this program is the planting of wildflowers on private property. It appears DOH may have strayed from one of their objectives by preserving and maintaining *non-native* wildflowers rather than native species.

Commissioner Douglas, responding to the question of invasive species, stated:

"With regard to the planting and cultivation of non-native plants that are considered to be invasive, here again there is little my agency can do. The West Virginia Department of Agriculture enforces the provisions of the West Virginia Noxious Weed Act, which prohibits the introduction and/or cultivation of certain plants known to be a detriment to agriculture or dangerous to human health. We recently amended the West Virginia Noxious Weed Act Rule to include additional plants, but we have no proof that any of the plants used in the Operation Wildlife Program are detrimental to West Virginia's agricultural interest or human health."

In a final statement from the Attorney General's office, Mr. McGraw concluded:

"By law there is nothing our office can do directly; however, we will continue to bring this issue to the attention of the agencies that hold jurisdiction over invasive species."

Six months of working with the Attorney General's office offered a poignant insight into the issue of invasive species in West Virginia. First and foremost, it has become obvious that there is a lack of knowledge of and information available to the various state agencies about the identity of native plants and who exactly has the responsibility of dealing with invasive species. There simply is no statewide policy of identifying and controlling native species. Each agency apparently has their own mandate and understanding of what they are required to do. Instead of a uniform policy, each agency works independently of each other.

What can be done to help alleviate this situation? Educating the general public *and* state agencies about the environmental threats posed by invasive species is of paramount importance. The landscape industry, nurseries, and garden centers must be involved in any discussion about eradicating invasive species. Many invasive species unfortunately have been introduced through the nursery trade. In fact, some of these species are still offered for sale. The sale of invasive species must be prohibited. Getting people to recognize and use native plants in the landscape can only be accomplished through educational programs. Any organization involved with native plants needs to take a proactive position by offering programs (workshops, lectures, and field trips), developing and distributing educational material, and inviting more of the public and private sector into their membership. Sadly, most of the general public remains unaware of the identity and diversity of native plants. Their first-hand experience with these plants all too often comes from seeing them grow along the roadside of interstate highways and country roads. Some will unwittingly dig these plants from the wild and attempt to grow them in their gardens. When the plants fail to survive the shock of being uprooted from their habitats, it leaves the mistaken impression with the gardener that native plants are too difficult or finicky to grow. It is so much easier and convenient, they perceive, to purchase non-native plants from local garden centers.

Tackling the issue of invasive species is a daunting undertaking. But every day that we delay only makes the task more formidable. If we think that we are overwhelmed today, imagine what it will be like in another decade. Stewardship—who will take responsibility? Native plants cannot speak for themselves.

SOME Thoughts About the Future 27

I am highly optimistic about the future of restoring and saving our natural areas. Organizations such as the Nature Conservancy, the Sierra Club, Wild Ones, and individual state native plant societies are educating the public about the myriad of benefits we derive from using native plants. And more and more books are being published about landscaping with native plants. This wave of enthusiasm must be sustained.

One area that we need to concentrate on involves the various state and federal agencies that are charged with the responsibility of preserving our natural areas. All too often, one agency is entirely unaware of the work being done by another agency. And it is even more frustrating to discover that some agencies ignore their mandates and foolishly use non-native and invasive species in their plantings.

Canadian Crown Vetch and non-native grass species are still used to plant along the sides of state highways. And yet in state and federal forests and parks, other agencies are fighting to eradicate the same species. We need to establish state guidelines that direct each and every agency about the ecological harm done by invasive species and to stipulate that native plants will be used in their place.

We need to have legislation passed that prohibits the use of any non-native species of plants on public property. Every tree, shrub, flower and grass planted should be indigenous to that state. Flowerbeds should be representative of the wealth of native species still growing in the wild. And educational material needs to be available to further the knowledge of native species.

Much has been accomplished, but there is still so much more that needs to be done. The future of our society rests in the hands of those who would nurture the natural wealth that has made this country what it is today.

Bibliographical Essay 28

Nothing heightens the senses more than walking through a natural area and coming upon a plant that you have not seen before. You immediately want to know the name of the plant and learn more about its growing habits. If you remembered to bring your camera, you take pictures of the flowers, leaves, and surrounding habitat. Hopefully, you also remembered to bring a notebook to describe where you found the plant, when you found it, and any other pertinent information. One can now begin the fascinating adventure of learning about native plants. It is time to take a trip to the local library and begin to familiarize yourself with the native plants of your region.

The literature about our native flora is as vast as the plants themselves. My journey began by reading Robert H. Mohlenbrock's *Where Have All the Wildflowers Gone? A Region by Region Guide* to *Threatened and Endangered U.S. Wildflowers*. With a deft hand, Mohlenbrock leads you through some of the richest and often unexplored areas in the United States. In addition to being an excellent field and reference guide, he describes in detail the fifty-five species on the federal and state lists of threatened plants at that time. Once you realize how many species are in jeopardy, you cannot wait to learn more about them and what can be done to help protect these valuable resources.

Learning to identify native plants can seem to be a daunting undertaking, especially for those who never took classes in botany. And being confronted with the Latin botanical names of the plants can strike fear in many. And yet learning to identify plants is one of the truly satisfying experiences as you begin your journey. Never having had a class in botany, I began identifying plants by using Roger Peterson's *A Field Guide to Wildflowers of the Northeastern and North-central North America*. Unlike the botanies written by trained professionals, this guide makes it easy to recognize flowers by sight. Focusing on the color of the bloom, the general shape or structure of the plant, and differences between species, the reader can begin to recognize through visual impressions the flowers in the field. Divided into six main color sections, you can quickly narrow your search. Nearly thirteen hundred species of wildflowers, shrubs, and woody vines are included. Nevertheless, the novice is still left to thumb through the pages hoping to find their plant.

Lawrence Newcomb's *Newcomb's Wildflower Guide: An Ingenious New Key System for Quick, Positive Field Identification of the Wildflowers, Flowering Shrubs and Vines of Northeastern and North-central North America* allows both amateurs and experts to identify almost any wildflower by focusing on the natural structural features that are easily visible to the untrained eye. You simply ask five questions that are related to the type of plant and the structure of its petals and leaves. You are then directed to the text page where the plant can be found. Excellent color and black and white illustrations confirm the identity of the plant.

As you become more proficient in identifying plants, you will want more detailed and comprehensive books about the plants in your part of the country. Dennis Horn's and Tavia Cathcart's *Wildflowers of Tennessee the Ohio Valley and the Southern Appalachians* offers coverage of a sixteen state region that includes Tennessee, the entire Ohio Valley, all of the central and southern Appalachians, the Piedmont, the mid-Mississippi Valley, and the Ozarks. More than twelve hundred species are described in detail. Jack B. Carman's *Wildflowers of Tennessee* is the first comprehensive, statewide, and full-color guide of the wildflowers of the Volunteer State. More than eleven hundred vascular plants and showier non-woody plants are thoroughly covered. Kay Yatskievych's *Field Guide to Indiana Wildflowers* provides information and illustrations for more than forty percent of the state's wildflowers not included in other field guides. More than fifteen hundred species are discussed and color photographs accompany the text. Richard M. Smith's *Wildflowers of the Southern Mountains* is a complete and non-technical field guide to the wildflowers of the southern mountains, stretching from Pennsylvania into northern Georgia. Despite the apparent overlap in coverage, each of the volumes contains information about species not found in the other books. One has at their fingertips virtually an encyclopedia of information about the wildflowers of a highly diverse part of the eastern United States.

As your knowledge and skills with identifying this native flora increases, so will your need for more detailed information. It is time to consult the professional floras of the eastern United States. Being a trained historian, I am interested not only in the plants themselves but in the men and women who began compiling these floras. It is interesting to found out when plants were first discovered, who discovered and named them, and how these taxonomic names have changed through time. I like to begin with Nathaniel Lord Britton and Addison Brown's *Illustrated Flora of the Northern United States, Canada and the British Possessions*. It was the first complete illustrated flora of this country. There have been several editions of this work, and each one added more species and illustrations. Henry A. Gleason thoroughly revised this work in 1952 as *The New Britton and Brown Illustrated Flora*. To make the work more readily accessible in the field, Arthur Cronquist finished Gleason's work. It appeared as *Manual of Vascular Plants of Northeastern United States and Adjacent Canada*. It remains indispensable for those doing fieldwork.

Asa Gray, curator of the Gray Herbarium at Harvard University, issued the first edition of his *Manual of the Botany of the Northern United States* in 1848. Numerous editions followed. In 1950, Merritt Lyndon Fernald, drawing upon the expertise and fieldwork of numerous scholars, prepared the eighth edition which was largely rewritten and expanded. John Kunkel Small, an American botanist, was the first curator of museums at the New York Botanical Gardens. He later became head curator. In 1903, Small published the *Flora of the Southeastern United States*, which was his dissertation at Columbia University. It was revised and published again in 1913 and 1933 as *Manual of the Southeastern Flora*. It remains one of the best references for the flora of the south. Small was one of the first botanists to thoroughly explore the flora of Florida. He returned numerous times between 1903 and 1931 to collect plants. His book, *From Eden to*

Sahara—Florida's Tragedy, received acclaim in 1929 for documenting the ecological destruction of south Florida's botanical resources that he had observed up to that time.

Botanists have been exploring the diverse geography of West Virginia in their search for new plants for over a century. In 1889, Charles Frederick Millspaugh began work as a botanist with the West Virginia University Agricultural Experimental Station. It immediately became apparent that little botanical work had previously been done. Millspaugh began collecting plants, hoping to have one specimen for each species in the state. The following year, he organized a botanical expedition by team and wagon with the station's entomologists. The trip covered 376 miles and resulted in the addition of many species to his herbarium. These early explorations during his three-year stay in West Virginia culminated in his *Flora of West Virginia* which was published in 1892. Beginning in the 1920s, P. D. Strasbaugh and Earl L. Core spent a quarter of a century continuing to collect and study the native flora of West Virginia. Their publication, the *Flora of West Virginia*, remains the standard reference for the Mountain State.

In addition to the tremendous number of wildflower species, there is also a wealth of native grasses, sedges, and rushes, but these are often overlooked. Even when we come across these plants during our jaunts, their identity often remains a mystery. For some folks, all grasses look alike. For others, the challenge of learning how to identify these plants is too much of an obstacle. But like our efforts to learn how to identify the wildflowers, we have to begin with the basics. A good start is to read Agnes Chase's *First Book of Grasses: The Structure of Grasses Explained for Beginners*; Edward Knobel's *Field Guide to the Grasses, Sedges and Rushes of the United States*; and

H. D. Harrington's *How to Identify Grasses & Grasslike Plants*. As you begin to learn the terminology describing the various parts of the grasses, you will be on your way to identifying.

Lauren Brown's *Grasses: An Identification Guide*, the first and only popular guide to the identification of grasses, avoids the technicalities of many botanical guides and instead focuses on the color, shape, and texture of the plants. Richard W. Pohl's *How To Know the Grasses: Picture-Keys for Determining the Common and Important American Grasses with Suggestions and Aids for Their Study* offers a more advanced approach to identifying the plants in the wild.

Equally intimidating is the identification of the sedges. Andrew L. Hipp's *Field Guide to Wisconsin Sedges* helps the reader to recognize key structures needed to identify Carex species.

The definitive study of the grasses of the United States remains A. S. Hitchcock's *Manual of the Grasses of the United States*. Originally published in 1935, it was revised in 1951. For anyone interested in studying our native grasses, this monumental study should be a part of your reference library. Excellent studies also have been made of the native grasses of each state. In some instances, these manuals are an up-to-date compilation of all native (and naturalized) grasses known to occur in the respective states. Although many of them appeared about the time of the revised edition of Hitchcock's classic study, they provide a greater focus on the grasses of each state. I have found the following studies to be extremely useful: Norman C. Fassett's *Grasses of Wisconsin*; Clair L. Kucera's *The Grasses of Missouri*; H. L. Blomquist's *The Grasses of North Carolina*; E. L. Core's *West Virginia Grasses*; and Clara G. Weishaupt's *A Descriptive Key to the Grasses of Ohio Based on Vegetative Characters*. Unfortunately, no adequate study has been done about the use of native grasses in the landscape. All too often, native grasses are lumped into volumes that discuss ornamental grasses from other parts of the world. Even so, Rick Darke's *Ornamental Grasses* can offer useful suggestions about incorporating native grasses into your landscape.

During your walks in the field, you will also encounter trees, shrubs, and vines. Many of these woody plants make excellent additions to your landscape. Woody vines are rarely treated as a separate group in many of the guides, but Wilbur H. Duncan's *Woody Vines of the Southeastern United States* was an important step towards remedying this situation. Native trees and shrubs frequently display striking flowers, fruits, and foliage that enhance the texture and color of your landscape. Guy Sternburg's and Jim Wilson's *Landscaping with Native Trees* stress that native trees can be disease- and pest-resistant, tolerant of weather extremes, and highly versatile. In a very easy-to-read style, they discuss the advantages and disadvantages of nearly every species of tree native to eastern North America. Richard E. Bir's *Growing and Propagating Showy Native Woody Plants* identifies some of the showiest native plants of the eastern United States. Oscar W. Gupton's and Fred C. Swope's *Trees and Shrubs of Virginia* is a non-technical guide with color photographs that provide a habitat picture and close-up of flowers or fruit. Don Kurz's *Shrubs and Woody Vines of Missouri* gives information about native shrubs and vines that can be used in landscaping and for providing habitat for wildlife.

Learning how to identify plants found in the wild and studying how they have adapted to their habitats has been the beginning of your journey. It is time to think about designing and planting your own native gardens. Carolyn Summers' *Designing Gardens with the Flora of the American East* explains why we should garden with indigenous plants, suggests ornamental substitutes for common invasive species, and provides designs based on indigenous plant communities and traditional gardens. Samuel B. Jones' and Leonard E. Foote's *Gardening with Native Wild Flowers* was especially written for those who are interested in creating landscape settings using native flowers and ferns. They easily discredited the common notion that the cultivation of wildflowers can be a complicated and difficult endeavor.

More recently, attention has been given to landscaping on a more regional scale. These authors help you identify the plants of your state and then give you the necessary information to grow them in a natural garden. *Landscaping with Native Plants of Michigan* and *Landscaping with Native Plants of Minnesota*, both written by Lynn M. Steiner, can serve as excellent examples of what can be accomplished in neighboring states that share many of the same species and habitats. For anyone interested in habitat restoration, Donald J. Leopold's *Native Plants of the Northeast: A Guide for Gardening & Conservation* describes natural plant communities of eastern North America and provides a selection of the best plants for a variety of sites in the garden. Equally informative and a delight to read is Carolyn Harstad's *Go Native: Gardening with Native Plants and Wildflowers in the Lower Midwest*.

On any walk through the woods and fields, you will encounter what seems to be an endless array of plants. So it is with the literature of the native plants of the eastern United States. Before closing, I would like to note some of the books that have been particularly useful and enjoyable companions on my treks in search of plants. Perry K. Peskin, who I had the pleasure to meet, was a retired school teacher from Cleveland, Ohio. Perry spent more thirty years in search of rare and endangered plants during trips that took him into the wilds of

Canada and the mountains of Appalachia. His book, *The Search for Lost Habitats*, allows readers to close their eyes and imagine being along on the search.

I continue to be fascinated by the exploits of the early plant hunters. The difficulties they faced as they searched for seeds and plants of the New World are amazing. Even in more recent times, plant hunters have continued to seek out new species or rediscover those that have not been seen in many years. Cora Steyermark's *Behind the Scenes* is a wonderful and insightful account of the work of her husband, Julian Steyermark, as they explored every region of Missouri from the 1930s through the 1960s and gathered information for his monumental *Flora of Missouri*.

All journeys must come to an end. As you become more and more familiar with the native plants of your part of the country, take the time to explore the literature written by the men and women who spent their lives studying these plants with the hope that others would share their love and enthusiasm for a natural resource whose future has been at times in doubt. As your garden grows in size, let your reference library keep pace.

Solidago nemoralis

ABOUT THE AUTHOR

Dr. Frank W. Porter earned his PhD in Historical Geography at the University of Maryland and came to his appreciation of native plants by a circuitous route. He spent most of his childhood overseas as his father was transferred from one Army post to another, and he has vivid memories of fields of poppies in France, alpine plants in the Alps, and dark forests in Germany. After a career as a Native American scholar, Dr. Porter returned to southern Ohio and began growing plants as a hobby, which quickly turned into a vocation. Porterbrook Native Plants nursery, on the banks of the Ohio River in southeastern Ohio, is devoted to the native plants of southern Ohio, West Virginia, and Virginia. Dr. Porter continues to collect seeds from interesting and endangered species in the area, and friends and associates send him more seeds from plants in neighboring states. He provides native plants for gardeners and public gardens, which he also designs, throughout Ohio and West Virginia.

Frank Porter

FLOWER TABLES

**PLANTS THAT CAN GROW IN
DRY SOILS AND PART SHADE**
Aquilegia canadensis (Red Columbine)
Campanula divaricata (Small Bonny Bellflower)
Chrysogonum virginianum (Green and Gold)
Coreopsis verticillata (Whorled Tickseed)
Dodecatheon meadii (Pride of Ohio)
Elymus Hystrix (Eastern Bottlebrush Grass)
Eurybia divaricata (White Wood Aster)
Eurybia macrophylla (Largeleaf Aster)
Geranium maculatum (Spotted Geranium)
Heuchera villosa (Hairy Alumroot)
Iris cristata (Dwarf Crested Iris)
Mertensia virginica (Virginia Bluebells)
Monarda bradburiana (Eastern Beebalm)
Ruellia strepens (Limestone Wild Petunia)
Sedum ternatum (Woodland Stonecrop)
Silene stellata (Widowsfrill)
Smilacena racemosa (Solomon's Plume)
Solidago caesia (Wreath Goldenrod)
Symphyotrichum cordifolium (Heartleaved Aster)
Symphyotrichum patens (Late Purple Aster)
Thalictrum dioicum (Early Meadow Rue)
Thalictrum thalictroides (Rue Anemone)
Tradescantia ohiensis (Bluejacket)
Tradescantia virginiana (Virginia Spiderwort)

WETLAND PLANTS
Actea racemosa (Black Bugbane)
Alisma triviale (Northern Water Plantain)
Andropogon glomeratus (Bushy Bluestem)
Baptisia australis (Blue Wild Indigo)
Caltha palustris (Yellow Marsh Marigold)
Chelone glabra (White Turtlehead)
Chelone lyonii (Red Turtlehead)
Conoclinium coelestinum (Blue Mistflower)
Hydrophyllum canadense (Bluntleaf Waterleaf)
Hydrophyllum virginianum (Eastern Waterleaf)
Iris brevicaulis (Zigzag Iris)
Iris cristata (Dwarf Crested Iris)
Iris virginica var. schrevrii (Shreve's Iris)
Juncus torreyi (Torrey's Rush)
Liatris spicata (Dense Blazing Star)

Lilium superbum (Turk's Cap Lily)
Lobelia cardinalis (Cardinal Flower)
Lobelia siphilitica (Great Blue Lobelia)
Mitella diphylla (Twoleaf Miterwort)
Physostegia virginianum (Obedient Plant)
Polygonatum biflorum (Smooth Solomon's Seal)
Sanguisorba canadensis (Canada Burnet)
Sisyrinchium angustifolium (Narrow Blue-eyed Grass)
Stylophorum diphylllum (Celandine Poppy)
Tiarella cordifolia (Heartleaf Foamflower)
Veronicastrum virginicum (Culver's Root)

NATIVE PLANTS FOR ROCK GARDENS
Allium cernuum (Nodding Onion)
Antennaria neglecta (Field Pussytoes)
Antennaria virginica (Shale Barren Pussytoes)
Asarum canadense (Canadian Wildginger)
Campanula rotundifolia (Bluebell Bellflower)
Claytonia virginica (Virginia Springbeauty)
Draba ramissimosa (Branched Draba)
Eriogonum allenii (Shale Barren Buckwheat)
Hepatica nobilis var. obtusa (Roundlobe Hepatica)
Hepatica nobilis var. acuta (Sharplobe Hepatica)
Heuchera villosa (Hairy Alumroot)
Hexastylis arifolia (Littlebrownjug)
Hexastylis heterophylla (Variableleaf Hertleaf)
Houstonia longifolia (Longleaf Summer Bluet)
Hylotelephium telephioides (Allegheny Stonecrop)
Hypericum mutilum (Dwarf St. Johnswort)
Hypoxis hirsuta (Common Goldstar)
Ionactis linariifolius (Flaxleaf Whitetop Aster)
Iris cristata (Dwarf Crested Iris)
Iris lacustris (Dwarf Lake Iris)
Krigia biflora (Twoleaf Dwarfdandelion)
Krigia montana (Mountain Dwarfdandelion)
Liatris microcephela (Smallhead Blazing Star)
Liatris squarrosa (Scaly Blazing Star)
Minuartia michauxii var. michauxii
    (Michaux's Stitchwort)
Oxalis violacea (Violet Woodsorrel)
Penstemon canescens (Eastern Gray Beardtongue)
Pheneranthes teretifolius (Quill Fameflower)
Phlox stolonifera (Creeping Phlox)

Phlox latifolia (Wideflower Phlox)
Ruellia humilis (Fringeleaft Wild Petunia)
Salvia lyrata (Lyreleaf Sage)
Saxifraga virginiensis (Early Saxifrage)
Scutellaria ovata (Heartleaf Skullcap)
Scutellaria parvula (Small Skullcap)
Sedum glaucophyllum (Cliff Stonecrop)
Sedum ternatum (Woodland Stonecrop)
Silene caroliniana (Sticky Catchfly)
Silene virginica (Fire Pink)
Taenidia integerrima (Yellow Pimpernel)
Thalictrum thalictroides (Rue Anemone)
Viola pedata (Birdfoot Violet)
Zizia aptera (Meadow Zizia)

PLANTS THAT ATTRACT BUTTERFLIES

Actea racemosa (Black Bugbane)
Agastache nepetoides (Yellow Giant Hyssop)
Amsonia hubrichtii (Hubricht's Bluestar)
Amsonia illustis (Ozark Bluestar)
Amsonia tabernaemontana (Eastern Bluestar)
Antennaria virginica (Shale Barren Pussytoes)
Aruncus dioicus (Bride's Feathers)
Aster species
Ceanothus americanus (New Jersey Tea)
Chelone glabra (White Turtlehead)
Chrysopsis mariana (Maryland Goldenaster)
Coreopsis species
Echinacea species
Eryngium yuccifolium (Button Eryngo)
Gentiana species
Geranium maculatum (Spotted Geranium)
Helenium species
Heliopsis helianthoides (Smooth Oxeye)
Hibiscus species
Marshallia grandiflora (Monongahela Barbara's Buttons)
Liatris species
Lilium canadense (Canada Lily)
Lilium superbum (Turk's Cap Lily)
Lilium michiganense (Michigan Lily)
Lobelia cardinalis (Cardinal Flower)
Monarda species
Penstemon species
Phlox species
Physostegia virginiana (Obedient Plant)
Pycnanthemum species
Ratibida pinnata (Pinnate Prairie Coneflower)
Rudbeckia species
Scutellaria species

Silphium species
Solidago species
Tiarella cordifolia (Heartleaf Foamflower)
Verbena species
Vernonia gigantea (Giant Ironweed)
Viola species
Zizia species

SUGGESTED PLANT LIST FOR RAIN GARDENS

Wildflowers:
Baptisia australis (Blue Wild Indigo)
Chelone glabra (White Turtlehead)
Chelone obliqua (Red Turtlehead)
Eutrochium purpureum (Sweetscented Joe Pye Weed)
Iris versicolor (Harlequin Blueflag)
Iris fulva (Copper Iris)
Liatris spicata (Dense Blazing Star)
Lobelia cardinalis (Cardinal Flower)
Lobelia siphilitica (Great Blue Lobelia)
Oligoneuron ohioense (Ohio Goldenrod)
Physostegia virginiana (Obedient Plant)
Polemonium reptans (Greek Valerian)
Rudbeckia laciniata (Cutleaf Coneflower)
Silphium perfoliatum (Cup Plant)
Stylophorum diphyllum (Celandine Poppy)
Symphyotrichum novae-angliae (New England Aster)
Thalictrum pubescens (King of the Meadow)
Verbena hastata (Swamp Verbena)

Grasses and Grass-like Plants:
Andropogon gerardii (Big Bluestem)
Carex species (Sedges)
Elymus virginicus (Virginia Wildrye)
Glyceria canadensis (Rattlesnake Mannagrass)
Juncus effusus (Common Rush)
Juncus torreyi (Torrey's Rush)
Spartina pectinata (Prairie Cordgrass)

Ferns:
Matteuccia struthiopteris (Ostrich Fern)
Onoclea sensibilis (Sensitive Fern)
Osmunda cinnamomea (Cinnamon Fern)

Shrubs and Trees:
Acer rubrum (Red Maple)
Clethra alnifolia (Coastal Sweetpepperbush)
Cornus amomum (Silky Dogwood)
Ilex verticillata (Common Winterberry)
Photinia melanocarpa (Black Chokeberry)

REFERENCES

Hellander, Martha E. *The Wild Gardener: The Life and Selected Writings of Eloise Butler*. St. Cloud: North Star Press, 1992.

Kirkpatrick, Golda and Holzwarth, Charlene. *The Botanist and Her Muleskinner. Lilla Leach and John Roy Leach: Pioneer Botanists in the Siskiyou Mountains*. Portland, 1984.

Laycock, George. *The Richard and Lucille Durrell Edge of Appalachian Preserve System Adams County, Ohio*. Cincinnati: Cincinnati Museum Center, 2003.

Smith, Beatrice Scheer. *A Painted Herbarium: The Life and Art of Emily Hitchcock Terry (1838-1921)*. Minneapolis: University of Minnesota Press, 1992.

Stuckey, Ronald L. *E. Lucy Braun (1889-1971): Ohio's Foremost Woman Botanist. Her Studies of Prairies and Their Phytogeographical Relationships*. Columbus: RLS, 2001.

A Vision of Eden: The Life and Works of Marianne North. Exeter: Webb & Bower, 1980.

INDEX

A

Adlumia fungosa, 57-58
Agalinis auriculata, 133
Ailanthus altissima 18, 46, 140
Allium cernuum, 38
Antennaria plantaginifolia, 94
Antennaria virginica, 38, 94, 99
Arabis serotina, 124
Aristolochia macrophylla, 40-42, 58
Aristolochia tomentosa, 40-42, 57, 58
Asarum canadense, 43
Asclepias, pollination of, 78-79
Asclepias
  amplexicaulis, 82
  cinera, 82
  exaltata, 80
  incarnata, 79-80, 118
  lanceolata, 82
  longifolia, 82
  obovata, 83
  pedicellata, 83
  perennis, 82
  purpurascens, 82
  quadrifolia, 79, 82
  rubra, 82
  speciosa, 80
  syriaca, 79-80
  tuberosa, 79-80
  verticillata, 82
  viridiflora, 83
  viridis, 83
Athyrium filix-femina, 40

B

Bignonia capreolata, 58
Brachyelytrum erectum, 42
Braun, E. Lucy, 132-135
Bromus kalmii, 42
Bromus pubescens, 42
Butler, Eloise, 127-128

C

Callicarpa americana, 54
Calycanthus floridus, 54
Campanula rotundifolia, 99
Campsis radicans, 58
Carex
  glaucodea, 95
  grayii, 119
  juniperorum 133
  pensylvanica, 42
  platyphylla, 95
Ceanothus americanus, 54
Celastrus scandens, 59
Centrosema virginianum, 59
Cercis canadensis, 47
Chasmanthium latifolium, 42
Cheilanthes
  eatonii, 37
  lanosa 37
Chelone glabra, 118
Chionanthus virginicus, 48
Chrysogonum virginianum, 43
Cinna arundinacea, 42
Clematis
  albicoma, 123
  viorna, 59
  virginiana, 59
Clethra
  acuminata, 52
  alnifolia, 52
Cornus
  alternifolia, 47
  florida, 47
Cotinus obovatus, 48

D

Danthonia spicata, 42, 95
Dennstaedia punctiloba, 40
Diarrhena americana, 42
Diervilla sessilifolia, 53

Dodecatheon meadia, 33
Doellingeria umbellate, 75
Draba ramosissima, 98
Dryopteris goldiana, 40

E

Edge of Appalachia Preserve System, 11
Eleocharis
  obtusa, 104
  tenuis, 104
Epigaea repens, 43
Equisetum hyemale, 38-39, 104
Eriogonum allenii, 38
Euonymous atropurpurescens, 49, 55

F

fern allies, 40
ferns, 37-38, 40
Filipendua rubra, 118

G

Gelsemium sempervirens, 42, 60

H

Halesia carolina, 48
Hamamelis virginiana, 48
Heracleum manztegazzianum, 18, 140
Hexastylis arifolia var. arifolia, 43
Hexastylis shuttleworthii var. shuttleworthii, 43
Huperzia lucidula, 40
Hydrangea
  arborescens, 52
  quercifolia, 53
Hylotelephium telephioides, 93, 99
Hypericum
  kalmianum, 54
  prolificum, 54
hypertufa containers, 93, 96-99

I

Ionactis linariifolium, 75, 99

Itea virginica, 53
invasive species, 15-19
Iris
  cristata, 31, 94, 104
  fulva, 30, 38, 118
  lacustris, 38, 104
  verna, 94, 104
  virginica, 30, 38, 118
  virginica var. Schreberi, 118

J

Junceae, 28

K

Kate's Mountain, 10
Kate's Mountain clover, 123

L

Leach, Lilla, 126-127
Liatris
  aspera, 66
  cylindracea, 66
  hellerii, 66
  ligulistylis, 66
  microcephala, 64-65, 67
  mucronata, 66-67
  odoratissima, 67
  pycnostachya, 64
  scariosa var. nieuwlandii, 67
  spicata, 64, 118
  squarrosa, 66
Lindera benzoin, 54
Locke, Dr. John, 131
Lobelia
  appendiculata, 88
  appendiculata var. Gattingeri, 90
  boykinii, 89
  cardinalis, 85, 86-87
  dortmannna, 89
  elongata, 89
  floridana, 90
  glandulosa, 89
  homophylla, 90
  inflata, 86, 87
  kalmii, 89

nuttallii, 89
puberula, 87
siphilitica, 85, 86, 87, 118
spicata, 88
Lobelia toxicity, 86, 88
Lonicera sempervirens, 42, 60
Luzula
  acuminata, 95, 103
  multifora, 95, 103
Lycopodium obscurum, 40
Lynx Prairie, 134-135

M

manicured lawns, 6-7, 27
Matelea
  carolinensis, 60
  obliqua, 60
McGraw, Daryl, 136-137
Meehania cordata, 95, 99
Menispermum canadense, 60-61
Mertensia virginica, 38
Microstegia vimineum, 16, 140
Miscanthus sinensis, xvi-xvii, 16, 140
Mitchella repens, 43
Mitella diphylla, 39

N

native gardens, preparation of, 21-23
  problem spots, 29
native grasses, 25-28, 42-43
native groundcovers, 36-43
North American Rock Garden Society, 91
North, Marianne, 128-130

O

Oclemena acuminata, 75
Oenothera argicolla, 124
Oligoneuron
  bicolor, 71
  ohioense, 71
  rugosa var. rugosa
Osmunda
  cinnamomea, 40
  claytoniana, 40

P

Pachysandra procumbens, 43
Packera aurea, 39-40
Paronychia argyrocoma, 95
Passiflora
  incarnata, 62
  lutea, 62
Paulownia tomentosa, 47
Paxistima canbyi, 133
Phlox
  buckleyi, 123
  divaricata, 39
  stolonifera, 94-95
  subulata, 94
Photinia
  floribunda, 53
  melanocarpa, 53
Physocarpus opulifolius, 52-53
pollinators, 109-112
Polypodium virginianum, 37
prairies
  creating a, 125
  species to plant, 127
Ptelea trifoliata 48-49
Pyrus calleryana, 46

R

rain gardens, 113-119
rock gardens, 91-95
  construction of, 93

S

Saccharum alopecuroides, 43
Sedum
  glaucophyllum, 33-34, 93-94, 99
  nevii, 33
  ternatum, 34, 39, 93, 99
Sisyrinchium
  albidum, 102
  atlanticum, 102
  angustifolium, 102
  mucronatum, 102
Solidago
  caesia, 70
  canadensis, 71
  flexicaulis, 70

odora, 70
patula, 71
rigidum var. rigidum, 70
rugosum, 71
uliginosa, 71
ulmifolia, 70
Spiraea
  alba, 52
  tomentosa, 52
Staphylea trifolia, 49
Steele, Edward S., 121
Strophostyles umbellata, 62
Stylophorum diphyllum, 39
Symphyotrichum
  cordifolia, 74
  dumosum, 76
  ericoides, 74
  laeve, 76
  lateriflorum, 76
  lowrieanum, 74
  novae-angliae, 75
  oblongifolium, 76
  patens, 76
  prenanthoides, 74
  puniceum, 75
  sagittifolium, 74
  undulatum, 74

T

Taenidia
  integerrima, 35
  montana, 124
Teays River, 134
Terry, Emily Hitchcock, 127
Thelypteris noveborancensis, 40
Tiarella cordifolia, 39

U

Viburnum
  acerifolium, 53
  prunifolium, 53

W

Wisteria frutescens, 62

Waldsteinia fragarioides, 43
Woodwardia virginica, 39

X

Xerophyllum asphodeloides, 103
Xyris
  caroliniana, 102
  torta, 102